EVOLVING APPROACHES TO MANAGING MARINE RECREATIONAL FISHERIES

Edited by
Donald R. Leal
and
Vishwanie Maharaj

Published in partnership with
Property and Environment Research Center (PERC)

LEXINGTON BOOKS
A division of
ROWMAN & LITTLEFIELD PUBLISHERS, INC.
Lanham • Boulder • New York • Toronto • Plymouth, UK

Published in partnership with
Property and Environment Research Center (PERC)

LEXINGTON BOOKS

A division of Rowman & Littlefield Publishers, Inc.
A wholly owned subsidiary of The Rowman & Littlefield Publishing Group, Inc.
4501 Forbes Boulevard, Suite 200
Lanham, MD 20706

Estover Road
Plymouth PL6 7PY
United Kingdom

Copyright © 2009 by Property and Environment Research Center (PERC)

All rights reserved. No part of this publication may be reproduced, stored in a retrieval system, or transmitted in any form or by any means, electronic, mechanical, photocopying, recording, or otherwise, without the prior permission of the publisher.

British Library Cataloguing in Publication Information Available

Library of Congress Cataloging-in-Publication Data

Evolving approaches to managing marine recreational fisheries / edited by Donald R. Leal and Vishwanie Maharaj.
   p. cm.
Includes index.
ISBN-13: 978-0-7391-2802-2 (cloth : alk. paper)
ISBN-10: 0-7391-2802-7 (cloth : alk. paper)
ISBN-13: 978-0-7391-2803-9 (pbk. : alk. paper)
ISBN-10: 0-7391-2803-5 (pbk. : alk. paper)
eISBN-13: 978-0-7391-3018-6
eISBN-10: 0-7391-3018-8
  1. Fishery management. 2. Saltwater fishing 3. Right of property. I. Leal, Donald. II. Maharaj, Vishwanie.
SH328.E96 2009
338.4'779916—dc22
                                                   2008025235

Printed in the United States of America

∞™ The paper used in this publication meets the minimum requirements of American National Standard for Information Sciences—Permanence of Paper for Printed Library Materials, ANSI/NISO Z39.48–1992.

Contents

Acknowledgments		vii
Introduction *Donald R. Leal and Vishwanie Maharaj*		ix

Part I: Prospects for Recreational Fishing Rights

Chapter 1	Evolution of Property Rights: Lessons of Process and Potential for Pacific Northwest Recreational Fisheries *Susan S. Hanna*	3
Chapter 2	Recreational Fishing and New Zealand's Evolving Rights-Based System of Management *Basil M. H. Sharp*	23
Chapter 3	Can Transferable Rights Work in Recreational Fisheries? *Hwa Nyeon Kim, Richard T. Woodward, and Wade L. Griffin*	47

Part II: Integrating Management of Commercial and Recreational Fishing

Chapter 4	Allocation of Fishing Rights between Commercial and Recreational Fishers *Ragnar Arnason and Peter H. Pearse*	79
Chapter 5	Harmonizing Recreational and Commercial Fisheries: An Integrated Rights-Based Approach *Ragnar Arnason*	97

Part III: IFQs and the Commercial Charter Boat Sector

Chapter 6	Examining the Interface between Commercial Fishing and Sportfishing: A Property Rights Perspective *Keith R. Criddle*	123

| Chapter 7 | Sport Charter Boat Quota Systems: Predicting Impacts on Anglers and the Industry *James E. Wilen* | 149 |

Part IV: Management Strategies for Saltwater Anglers

| Chapter 8 | Fish Harvest Tags: An Attenuated Rights-Based Management Approach for Recreational Fisheries in the U.S. Gulf of Mexico *Robert J. Johnston, Daniel S. Holland, Vishwanie Maharaj, and Tammy Warner Campson* | 171 |
| Chapter 9 | Angling Management Organizations: Integrating the Recreational Sector into Fishery Management *Jon G. Sutinen and Robert J. Johnston* | 201 |

| Index | 231 |
| About the Contributors | 241 |

Acknowledgments

There were many who made publication of this volume possible. First and foremost, we thank the chapter authors who gave so willingly of their time, talent, and knowledge of marine fisheries. It has been a pleasure and a privilege working with them throughout the process.

Taking the original papers to a form suitable for publishing requires a special blend of proficiency, patience, and persistence. Fulfilling that role masterfully was Dianna Rienhart from the Property and Environment Research Center (PERC). Without her we would still be struggling to meet our deadlines. Thanks also to Diane Ehernberger of Big Sky Indexing for copyediting the manuscript and preparing the volume's index; to PERC staff members Renee Storm for helping with the formatting requirements, Michelle Johnson for re-creating tables and figures, and Laura Huggins and Mandy Bachelier for cover design.

This volume is the end product of a forum held in Big Sky, Montana, on October 5–8, 2006, jointly sponsored by PERC and the Texas Office of Environmental Defense Fund (EDF). We are grateful to Pamela Baker of EDF for not only spearheading the idea of such a forum, but for codirecting it with Donald Leal of PERC. Special mention must be made of Monica Guenther and Colleen Lane of PERC for providing invaluable logistical support to the forum and to the resulting book project.

Projects such as this require funding from people and foundations willing to make an investment in ideas. Fortunately, enough of them stepped up to make this project a reality. We thank Kristine Johnson of the Kingfisher Foundation, Barrett Walker of the Alex C. Walker Educational and Charitable Foundation, the Walton Family Foundation, and a generous anonymous donor.

Introduction

Donald R. Leal and Vishwanie Maharaj

Marine recreational fishing is a popular activity in the coastal regions of the United States enjoyed by more than thirteen million anglers annually, but along with such popularity come a number of issues that need to be addressed (NMFS 2007). For species of concern—those overfished or undergoing overfishing—the impact of recreational fishing can be especially significant. For example, recreational fishing accounts for 64 percent of the total annual harvest of red snapper and other species of concern in the U.S. Gulf of Mexico (Coleman et al. 2004).

For angling interests, fisheries regulation can be another issue. Prior regulatory tightening has not prevented recreational catches from often exceeding state-level targets in the East Coast summer flounder fishery—a fishery with preliminary 2006 recreational landings totaling 11.2 million pounds or 4.1 million fish (ASMFC 2007). To make up for the overages, significantly lower catch targets have been issued for New Jersey and six other states along the East Coast for 2008 (U.S. Department of Commerce 2008). Meeting the lower targets means tighter restrictions and inevitable disagreement over which restrictions should be tightened and by how much. Bay anglers spar with ocean anglers over minimum landing size, surf anglers are upset over possible fall closures, party boat operators disagree on bag limits, and bait and tackle shops are concerned with lower participation in the fishery. "We're dealing with geography and demographics among other things," says one angling spokesperson in New Jersey. "The needs of those in the fishery are so different in New Jersey" (quoted in Geiser 2008).

For fish stocks shared by recreational and commercial fishers, the problem of allocation and enforcement is another issue. Off Alaska, charter boat anglers face a possible reduction in the halibut bag limit—from two fish per day to one fish per day if their catch allocation is exceeded during the 2008 season (Associated Press 2008; NPFMC 2008). Commercial fishers complain that the "soft" recreational harvest limit for halibut is frequently exceeded. They have grown weary of recreation's increasing share of total halibut landings—from 2 percent of total landings in the 1970s to more than 18 percent in the 1990s (Criddle, this volume). But charter captains claim that any reduction in client bag limits would severely hurt their businesses.

For regional stocks threatened or endangered, fishing bans are another issue. Off California and most of Oregon, managers are set to ban all commercial ocean salmon fishing and drastically reduce recreational ocean salmon fishing in 2008 to protect the fabled Chinook salmon run in California's Sacramento River. For reasons other than fishing, the salmon run declined precipitously in 2007—from a previous average of 500,000 returning adults to 90,000 returning adults (McClure 2008; Koepf 2008). Oregon anglers complain that such severe restrictions prevent them from accessing healthy runs of salmon off coastal Oregon.

Three lessons should be noted here. One is that while marine recreational fishing is an activity carried out for sport and personal use, it is still susceptible to what Garret Hardin (1968) calls the "tragedy of the commons."[1] In this case, the tragedy manifests itself when the aggregate recreational catch often exceeds safe target levels or takes a growing share of the catch, resulting in growing conflicts with other users of the resource. The second lesson is that the traditional management approaches of reducing daily bag limits and seasons and increasing the minimum size for landing fish is often not enough to prevent overfishing (other than an outright ban on fishing), but they do generate angler discontent and lower economic benefits for those who service anglers. The third lesson is that implementing a one-size-fits-all set of restrictions over a large geographic area ignores widely varying preferences among angler populations and environmental conditions, resulting in additional angler dissatisfaction and further loss of economic benefits.

Is there an alternative approach to managing a recreational fishery—one in which the aggregate catch is effectively controlled, the fishing experience remains enjoyable, and the chance for user conflict is reduced? The answer is yes. Based on evidence from the areas of pollution control, inland hunting and fishing programs, and commercial fishing, market-tradable user rights can prevent overuse of a commons without having to resort to ever-tighter fishing restrictions. Moreover, they provide anglers and those who service them—charter boat operators, fishing guides, bait and tackle shop owners—with longer seasons, more flexible harvests, and the means to trade for additional allocations with other resource users. Organized in four parts, this volume investigates such alternatives as well as the theoretical and practical aspects of implementing transferable user rights in some of today's recreational fisheries.

In part I, the prospects for recreational fishing rights are explored in the volume's first three chapters. Susan S. Hanna (chapter 1) points out that marine recreational fishing in the United States is experiencing some of the same problems that led to adoption of rights-based fishing approaches in commercial fishing. Dr. Hanna then explores the prospects for rights-based fishing systems in three recreational fisheries in the Pacific Northwest. All three are experiencing reduced fishing opportunities and growing conflicts with their commercial counterparts. Recreational interests, however, have yet to express interest in rights-based approaches. Dr. Hanna finds that the drivers of institutional change exhibited in three commercial fisheries she examined are "only thinly in place" in these fisheries. Moreover, "social learning" about the benefits of rights-based approaches has not yet taken place in the recreational fisheries. Hence, a key initial step for institutional change in these fisheries would be to educate anglers and recreational managers on the opportunities under rights-based fishing.

In chapter 2, Basil M. H. Sharp explores the issue of integrating marine recreational fishing into New Zealand's commercial management system, a system based on individual transferable quotas (ITQs). Dr. Sharp points out that the process of allocating shares of the total allowable catch (TAC) between commercial and recreational interests is fundamentally flawed. Poor information on the value of recreational fishing and ineffective control of recreational harvests hamper the process. New Zealand's snapper and kahawai fisheries serve as illustrations. Both fisheries have been subjected to litigation following commercial and recreational harvest allocations by the Minister of Fisheries. In contrast, the South Island scallop fishery shows rare cooperation among commercial and recreational fishers. Looking ahead there are signs that recreational charter fishing operators might have to hold ITQs, or at least report their harvests to managers. Licensing of recreational fishers might also become a reality. A truly integrated rights-based approach to managing commercial and recreational fishing, however, requires collaboration between the commercial and recreational sectors as well as legislation that defines recreational rights for fishing organizations and empowers them to exclude nonmembers from designated fishing areas.

Hwa Nyeon Kim, Richard T. Woodward, and Wade L. Griffin (chapter 3) draw on lessons learned from application of transferable rights in pollution, inland fishing and hunting programs, and marine commercial fishing to determine whether transferable rights (TRs) would work in marine recreational fishing. The answer is yes if the program is designed in a way that makes TRs both practical and effective to use. A practical program is one that can easily be monitored and enforced, with the transaction costs for allocating and transferring rights kept to a minimum. To be effective such a program must ultimately control fish mortality while generating sufficient information for managers to gauge the impacts of fishing, the number of participants, and the extent of their fishing efforts. The authors see rights denominated in fish or pounds of fish as more precise in terms of controlling the catch but not practical in terms of monitoring and enforcement in a fishery with a large number of participants. The authors propose an alternative in which anglers, for-hire charters, or fishing groups bid for access days to a fishery and that these access days are transferable. Since the rights are denominated in fishing days, bag limits would be retained to prevent excessive catch.

Avoiding conflict and achieving efficient allocation of the catch among different users is the topic of part II. Ragnar Arnason and Peter H. Pearse (chapter 4) demonstrate that market trading of shares of the catch between commercial and recreational interests is the best way to ensure optimum allocation and avoid controversy. However, the challenges to reaching market capability are formidable because recreational fishers are "typically disparate" with "widely ranging values" for the fish. The authors review how such challenges are being met for the commercial and recreational sectors in Canada's Pacific halibut fishery.

The use of ITQs in commercial fisheries provides a way to control the catch more precisely while giving fishers more operational freedom to improve their bottom line. Given such benefits, could an ITQ system be used to better coordinate commercial and recreational use of a fish stock? In chapter 5, Ragnar Arnason investigates this question analytically and finds that an integrated ITQ system (i.e., one encompassing both recreational and commercial fishers) is indeed capable. Of course, such a result "assumes perfect and costless enforcement." In fisheries dominated by tens of thousands of anglers dispersed over a large section of the United States, enforcement of ITQs on recreational fishers can be "prohibitively costly" and modified approaches become necessary. Modifications are also needed when considering conservation interests because the benefits of a larger fish stock is not a private but a public good, argues Dr. Arnason.

Given the commercial aspect of the charter-based sport fishery, ITQs appear to be well suited as a management alternative—the topic of part III. In chapter 6, Keith R. Criddle uses a conceptual analysis to determine the effects of changes in commercial and charter-based sportfishing demands and fish abundance under four alternative management regimes. In one finding, an ITQ managed commercial fishery "helps mitigate the adverse impacts of increased catch allocations to the sport fishery." In another finding, when commercial and charter-based sport fisheries are under an integrated ITQ system, "the effects of fluctuations in stock abundance are immediately distributed across both fisheries." Dr. Criddle then uses an empirically based simulation to show the importance of an integrated system for managing commercial and sportfishing. An integrated system for managing commercial and recreational fishing of the same fish stock produces higher overall net benefits than one where commercial and recreational fishing are managed separately. Dr. Criddle cautions, however, that even this system does not include all interests in the fishery. The challenge is to devise a rights-based approach that includes commercial, sport, and other use and nonuse values.

Proposals for ITQs in the charter-based sport fishery have created concerns among angling interests about the possible impacts. Using a simple framework for analysis James E. Wilen (chapter 7) contrasts the fishery under status quo management with one under an ITQ program. Like commercial fisheries, a sport fishery under open access is prone to too many boats and fewer fish, resulting in higher operator costs. The consequence for client-anglers is higher trip prices to cover the inflated operating costs. Two typical regulatory responses to fewer fish is to shorten the season and/or to reduce the daily bag limit, but such measures can lead to lower angler satisfaction and lower trip demand for charter operators. By reducing fishing excesses an ITQ program lowers charter operating costs. Moreover, individual quotas for vessel operators allow greater flexibility to meet the varying demands among anglers. But there is an additional impact if the program evolves into one that charges anglers extra fees for harvesting fish. Such fees will probably affect resident and nonresident anglers differently in terms of willingness to pay.

A sustainable management system for the large and complex Gulf of Mexico reef fish fishery must control the recreational harvest, mitigate conflicts with other stakeholders, generate high quality data, and allow enough flexibility to meet the widely varying tastes of anglers spread over a large region. Strategies to address some or all of these requirements are the topics of part IV. Robert J. Johnston, Daniel S. Holland, Vishwanie Maharaj, and Tammy Warner Campson (chapter 8) examine hunting and fishing tag programs used in the United States and abroad to determine how such an approach might be applied to the Gulf snapper fishery and other similar fisheries. The authors see the use of a limited overall number of fish tags corresponding to the overall catch desired by managers as a more effective way of controlling the recreational harvest and generating more reliable data than relying strictly on traditional restrictions and phone surveys. Based on the programs examined, a fish tag program would appear to be more acceptable to anglers than one based on ITQs—at least in the short run.

A fish tag program is not a full-fledged durable rights system, and hence it is not the best choice for integrating management of commercial and recreational fishers or for accounting for other use and nonuse interests. Standard ITQ programs might seem to be one answer, but a large number of anglers complicates monitoring and enforcement. Jon G. Sutinen and Robert J. Johnston (chapter 9) offer a novel approach called "angling management organizations," or AMOs. Each AMO is a logical extension of anglers who share a common angling preference or other common trait and could represent their interest in bargaining with other AMOs, commercial fishers, or other stakeholders over shares of the catch. An AMO is also responsible for making sure that its allocated share of the total allowable catch is not exceeded by anglers within its jurisdiction. By assuming management responsibility for monitoring and enforcement of the AMO's share of the catch, management costs are lower than under an approach based on a standard ITQ system.

The chapters in this volume are designed to stimulate thinking on alternatives to managing marine recreational fisheries and the conditions necessary for constructive institutional change. Under existing institutions, fishery agencies mainly monitor and enforce gear and method restrictions, bag limits, seasonal closures, and minimum size limits, but they are severely limited in effectively monitoring or controlling the extent of participation and total harvests in a fishery. In addition, these agencies lack the necessary information and market tools for reconciling catch allocations efficiently between commercial, recreational, and other noncommercial interests in marine fish stocks. We hope that the chapters herein provide the brick-and-mortar ingredients for management reform of recreational fisheries—one that allows anglers greater enjoyment of their sport, for-hire sector businesses more flexibility and increased profits, and managers better control of the catch at lower costs.

Note

1. For a classic article on the fishery as a commons, see H. Scott Gordon (1954). See also Colin W. Clark (1981).

References

Associated Press. 2008. Halibut Charters May Lease Commercial Quotas. *Anchorage Daily News*. April 10. (Available at http://www.adn.com/money/story/371357.html.)
Atlantic States Marine Fisheries Commission (ASMFC). 2007. Species Profile: Summer Flounder. *ASMFC Fisheries Focus* 16(2): 2. (Available at http://www.asmfc.org/speciesDocuments/sfScupBSB/summerflounder/sFlounderProfile.pdf.)

Clark, Colin W. 1981. Profit Maximization and the Extinction of Animal Species. *Journal of Political Economy* 81(4): 950–61.

Coleman, Felicia D., Will F. Figuera, Jeffery S. Ueland, and Larry B. Crowder. 2004. The Impact of United States Recreational Fisheries on Marine Fish Populations. *Science* 305(5692): 1958–60.

Criddle, Keith R. 2008. Examining the Interface between Commercial Fishing and Sportfishing: A Property Rights Perspective. This volume.

Geiser, John. 2008. No Good Option among Summer Flounder Proposals. *Asbury Park Press*. February 29. (Available at http://www.app.com/apps/pbcs.dll/article?AID=/20080229/SPORTS06/802290444.)

Gordon, H. Scott. 1954. The Economic Theory of a Common Property Resource: The Fishery. *Journal of the Political Economy* 62: 124–42.

Hardin, Garrett. 1968. The Tragedy of the Commons. *Science* 162: 1243–48.

Koepf, Enrie. 2008. Where Did the Sacramento Salmon Go? *Pacific Fishing* April: 30. (Available at http://www.pacificfishing.com/art/PF_April08_Sacramento_Salmon.pdf.)

McClure, Robert. 2008. Salmon "Emergency" Spawns New Limits. *Seattle Post-Intelligencer*, April 11. (Available at http://seattlepi.nwsource.com/local/358664_salmon11.html.)

National Marine Fisheries Service (NMFS). 2007. *By the Numbers: Saltwater Fishing Facts and Figures 2006.* Silver Spring, MD: U.S. Department of Commerce, National Oceanic and Atmospheric Administration, NMFS, September. (Available at http://www.nmfs.noaa.gov/sfa/PartnershipsCommunications/recfish/BytheNumbers2006.pdf.)

North Pacific Fishery Management Council (NPFMC). 2008. *April 7, 2008 Council Action on Charter Halibut Catch Sharing Plan for Area 2C and Area 3A.* (Available at http://www.fakr.noaa.gov/npfmc/current_issues/halibut_issues/HalibutCharterMotion408.pdf.)

U.S. Department of Commerce, National Oceanic and Atmospheric Administration (NOAA), National Marine Fisheries Service (NMFS). 2008. Proposed Rules: Fisheries of the Northeastern United States; Recreational Management Measures for the Summer Flounder, Scup, and Black Sea Bass Fisheries; Fishing Year 2008. 50 CFR Part 648. *Federal Register* 73, no. 56 (March 21, 2008): 15111–18.

Part I
PROSPECTS FOR RECREATIONAL FISHING RIGHTS

Chapter 1

Evolution of Property Rights: Lessons of Process and Potential for Pacific Northwest Recreational Fisheries

Susan S. Hanna

The history of fishery management is one of long periods of incremental adjustment punctuated by short bursts of major change. The history involves various expressions of property rights, whose adoption is shaped by incentives, costs, and social values. The extended evolution of property rights in commercial fisheries is well documented (Iudicello, Weber, and Wieland 1999; Hannesson 2004; Leal 2005). At its essence it is a story of regulatory attempts to align the private incentives of fishing businesses with social objectives of conservation and sustainability. The same evolution in recreational fisheries has been shorter, slower, and less well documented. However, the fully exploited state of many marine fish stocks combined with growing demand for recreational use draws new attention to the need to address the role of recreational fisheries in achieving sustainable fisheries. Increasing U.S. coastal populations, increasing recreational demand, and growing conflicts among commercial and recreational uses all point to the need to explicitly address the question of management control over recreational fisheries. As with commercial fisheries, this question involves aligning public and private incentives and leads naturally to considerations of property rights.

In this chapter, I address the question of property rights for marine recreational fisheries in the Pacific Northwest. I examine recreational property

---

Research for this chapter was supported by Hatch Project Funds at the Coastal Oregon Marine Experiment Station (COMES), Project 605: Improving Marine Fisheries Management and Policy in Oregon.

rights in the larger fishery context, looking at the roles played by context and process in property rights evolution. I first describe recreational fisheries management in the region in terms of its objectives, regulatory controls, and emerging change. Next, I summarize the general theoretical understanding of the drivers of change in property rights. I then present case studies of change in three Pacific Northwest commercial fisheries. Against this background, I address the question of property rights in three regional recreational fisheries with allocation problems where the potential exists for rights-based management.

Pacific Recreational Fishery Management

The National Marine Fisheries Service's (NMFS) *Fisheries of the United States 2004* categorizes recreational catch data by species, location, and mode of fishing (NMFS 2005b). Locations are on-shore, state waters (in the Pacific out to three miles from shore) and federal waters (located further offshore from three to two hundred miles from shore). Modes of fishing are fishing from shore, by private boat, and by party boat. The mode accounting for the largest proportion of 2004 total U.S. recreational catch was private boat at 71 percent, followed by charter boat at 16 percent, and fishing from shore at 13 percent (NMFS 2005b, 31).

Pacific marine recreational fisheries are managed by the respective states in collaboration with the Pacific Fishery Management Council (PFMC). Recreational licenses are mandatory but not limited. Recreational catches are regulated using detailed combinations of restrictions such as size limits for landing fish, daily bag limits, season length, closed fishing areas, gear restrictions (e.g., fishing for Oregon sport salmon requires use of single point, single shank, barbless hooks), and methods of fishing designed to minimize incidental catch of non-targeted species (Oregon Department of Fish and Wildlife [ODFW] 2006).

This typical suite of management measures represents a layering of different restrictions over time as managers have attempted to limit total recreational catch while maximizing recreational fishing access. Most are input controls and only indirectly influence output. Bag limits are an attempt to control total output, but the absence of control over the number of anglers or trips (except indirectly through season length) makes the bag limit a crude approach that requires in-season monitoring to stay within the targeted overall recreational catch. The absence of limits to numbers of anglers or numbers of trips has, in the face of increasing recreational use, resulted in the predictable shrinking of bag limits and season lengths (ODFW 2006).

The trend in more stringent restrictions has been exacerbated by the overfished status of important groundfish stocks as well as special protections on several Pacific salmon stocks under the Endangered Species Act (ESA). The tightening of recreational management controls has brought to the fore unresolved allocation conflicts with the commercial sector. As Sutinen and Johnston (this volume) noted, these trends reflect the logical flaw of attempting to manage recreational fisheries without full controls over the overall recreational catch.

The absence of full catch accounting adds to the problem of controlling the total recreational fishery impact on fish stocks. Pacific recreational fisheries, like others nationwide, are characterized by large numbers of anglers landing fish in many locations, making catch and effort data difficult and expensive to obtain. Since 1979, the NMFS has sampled recreational fishery activities through the Marine Recreational Fisheries Statistics Survey (MRFSS), a telephone survey of coastal county households combined with an intercept survey of angler trips (NMFS 2005b). Oregon and Washington do not participate in the MRFSS but instead conduct their own ocean boat surveys to estimate catch and effort data. Both state and federal surveys indicate that recreational catch, while overall a small proportion of total U.S. catch, can nevertheless cause substantial impacts on certain stocks of popular recreational species (NMFS 2005b).

The question of what agencies are trying to accomplish with recreational fishery management is interesting and complex. Anglers are not a homogeneous group but instead have motivations that vary widely, and include sport, catch, experience and relaxation (Iverson 1996). Federal management institutions necessarily aspire to general outcomes. The 1995 Executive Order 12962 directed federal agencies to work together to "conserve, restore and enhance aquatic systems to provide for increased recreational fishing opportunities nationwide" (NMFS 2005a). This type of general goal for marine recreational fisheries is repeated in state and federal statute and administrative codes. However, the specific objectives under which recreational fisheries are managed vary across and within regions and may be subject to different interpretations over time.

Iverson (1996) cites a 1978 survey of recreational fisheries management goals that shows yield-oriented management goals (e.g., maximum sustainable yield [MSY], maximum catch, and trophy fisheries) ranked significantly higher than angler-oriented goals (e.g., maximum angler trips, maximum angler days, and minimized angler crowding) in state recreational fisheries programs. However, writing two years earlier, Larkin (1977) offers a contrasting view by defining maximum yield (benefit) in recreational fisheries as the greatest personal satisfaction obtained from using the least efficient gear with the greatest skill to produce the smallest catch.

Recent experience in the Pacific region with curtailed sportfishing seasons for groundfish and salmon suggests that angler-oriented goals such as angler days are a priority, reflected by state and federal management attempts to provide opportunities for the fishing experience even in times of biological scarcity (PFMC 2006b; ODFW 2006).[1]

Recreational management objectives are complicated because they attempt to maximize all components of fishing satisfaction. Anderson (1984, 19–20) notes that a major difficulty for a management body is to deconstruct the recreational fishing experience into its component parts to, for example, be able to understand how fishing satisfaction is related to the size of the catch versus other experiential components of a fishing trip. This is especially problematic when managing both commercial and recreational sectors, where pecuniary externalities exist between sectors. An optimal level of regulation will consider the contribution to total fishery value generated by each of the sectors and will adjust fishing opportunities across sectors accordingly. Determining such a level and making requisite adjustments in response to changing economic and biological conditions are problematic when relying strictly on political processes (Criddle, this volume).

Managers are continually faced with the need to allocate limited allowable catch across commercial and recreational sectors. The allocation requires a reasonable understanding of the value generated by each sector, but estimations of value are often incomplete, leaving the valuation process to the assertions of interest groups. Edwards (1991) notes that both recreational and commercial interests often exploit the absence of analysis by using arguments designed to achieve favorable allocations (e.g., the "revenues argument" based on a comparison of commercial gross revenues to angler expenditures to justify a sector allocation, or the "all-or-nothing" argument for an allocation of the total allowable catch based on comparisons of the economic value of the fishery in each sector). Having to choose among these interest-group arguments is not the only option for managers. As long as net economic values are being estimated following standard economic procedures (Edwards 1990 and 1991; Hundloe 2002), questions about whether or not a reallocation from one fishing sector to another increases fishery value can be evaluated in a systematic manner (Blamey 2002; Arnason, this volume).

However, the chronic paucity of recreational data combined with its large nonmarket component make valuing recreational fisheries, particularly in comparison to commercial fisheries, problematic (Hanna et al. 2006). The continuing challenge of recreational fishery or recreational-commercial valuation is made necessary by the failure to resolve the question of rights in the fishery. Without marketable rights, managers continue to be responsible for figuring out how to allocate fish stocks to highest value uses. Within commercial fisheries the value of rights-based management is now widely accepted even if there is disagreement about the best form. Within recreational fisheries rights are still a disputable concept. Sutinen and Johnston (this volume) list strong angling rights as a principle on which integrated recreational fishery management must be based. The growing political power of recreational fisheries threatens continual erosion of commercial fishery allocations (De Alessi 1998), and it makes stable joint management of commercial and recreational fisheries a formidable task.

Institutional Change and Property Rights

Adoption of property rights in fisheries is a process of institutional change involving the economics of incentives, costs, and opportunities. When property rights are poorly specified, the performance of institutions is plagued by a number of well described, efficiency-reducing incentive problems. These problems increase transaction costs and limit performance effectiveness by adding to uncertainties and shortening time horizons (Hanna 2006).

Power Ambiguity: Contestation over decision-making authority (Arrow 1974) has troubled the United States fishery management system since its inception. Although the relative roles and responsibilities at different levels of management hierarchy are detailed in law and implementing regulations, authority is challenged and power struggles persist.

Failure to Make Credible Commitments: Credible commitments exist when what is promised is reliably delivered (Williamson 1985). If property rights over ocean resources are absent, and if the management environment is uncertain or unstable, managers may be prevented from making either commitments or threats with credibility.

Low-intensive Incentives: Accountability is missing when the connection between a decision and its consequences is missing (Williamson 1985). The lack of direct accountability for management outcomes has been named as a problem facing fishery management.

Moral Hazard: Moral hazard exists when actions are unobservable and create the conditions for shirking (Eggertsson 1990). Although a number of elements of the "notice and comment" fishery management process are designed to promote transparency in decision making, complex decision and allocation rules create many obstacles to transparency.

Bounded Rationality: Bounded rationality is behavior that intends to be rational but is limited by uncertainty and inconsistency. Time horizons shorten, leading people to more strongly value present certain benefits over future uncertain benefits (Williamson 1985).

Truncated Learning: Learning-by-doing can be a way for organizations to increase proficiency, adapt to changing circumstances, and reduce costs, but opportunities for learning and adaptation can be limited by uncertain decision environments and strategic information shaping (Tirole 1995).

All these attributes of the incentive environment—power ambiguity, low-intensive incentives, moral hazard, bounded rationality, a lack of credible commitment, and truncated learning—create problems for fishery management. All complicate the application of knowledge in management and keep the private incentives of decision-makers and other management participants from being fully aligned with public objectives. They also create costs, which, in turn, generate pressures from various interests for institutional change to alleviate these costs.

North (1990) addresses institutional change in terms of the interaction of rights and constraints. Constraints that define the opportunity set for decision making are both formal (the hierarchy of rules) and informal (reputation, conventions, and standards of conduct). Formal constraints are backed by the authority of law. As such, only when it is in the interest of those with sufficient bargaining strength to alter rules will there be major changes in the formal institutional framework. Informal constraints have their own authority created by acceptability and effectiveness that can endow them with great "survival ability" (North 1990, 83). The "interest-group theory" of property rights sees property rights as the outcome of rent-seeking interactions among interest groups within the context of formal and informal constraints (Eggertsson 1990, 275).

Costs are generated by transactions that take place under an existing institutional structure and are also, in the case of changing property rights, created by the transformation to a new structure. The types and levels of these costs are influenced by elements of the fishery context. Costs of information gathering, monitoring, and enforcement are all affected by the combinations of formal and informal constraints. These constraints, and the nature of the transactions, affect both the magnitude and the distribution of institutional costs (Hanna 1995b).

An institutional equilibrium exists when the combination of rights and constraints create relative costs and benefits that do not make it worthwhile to change. It is only when changes in relative costs or benefits (economic or political) create opportunities for some to improve their situation that incentives for change are generated (North 1990). Change may be generated implicitly through neglect and erosion or explicitly through concerted action. It may be gradualistic and incremental, maintaining institutional stability, or rapid and quantum, replacing one institutional framework with another (Gersick 1991).

Williamson (1994) describes any mode of governance as a syndrome of attributes, so the move from one to another, such as through the adoption of property rights, involves trade-offs within that attribute mix. The trade-offs alter the relative transaction costs to individuals or groups and affect their incentives to support or oppose the change. Williamson's "fundamental transformation" (1985) describes how, through changing transaction costs and development of specific assets, a governance negotiation can change from one characterized by many individual competitors (before an agreement is reached) to one of contractual asymmetry (after the agreement). It is an illustration within the fishery management context of the importance of path dependence and how history matters to the balance of power in changing property rights.

Hirschman (1992) examines the influence of loyalty, voice, and exit in promoting institutional change. When transaction costs rise, loyalty to an existing institutional arrangement becomes strained. The use of "voice" by participants in the arrangement can alert management to these costs and related problems. Depending on the type of arrangement in place, and in particular the combination of rights and informal constraints under which it operates, voice may be costly or even dangerous to reputation or group cohesiveness. If for these reasons voice cannot be used to implement cost-reducing changes, participants may opt to use "exit," or withdrawal from the management arrangement. Exit is a blunt communication of failure to managers. It implies that alternatives exist for those who leave, and it may be one way to force institutional change.

The economics of organization and institutions can be understood through the lens of various units of analysis: the types of decision, the individual, the industry, the degree of ownership, or the transaction (Williamson 1996). The ownership lens—the form and distribution of property rights—integrates all other units by shaping the incentives and behavior through which they interact. The general theoretical findings about organizations and institutions are well illustrated by the paths of changing property rights in fisheries.

Paths of Changing Property Rights: Pacific Commercial Fishery Examples

Three cases of changing property rights in Pacific Northwest commercial fisheries illustrate the dynamics described in the general literature.

Property Rights Progress: Pacific Groundfish License Limitation

The commercial multispecies groundfish fishery off Washington, Oregon, and California expanded rapidly after the 1977 implementation of the Fishery

Conservation and Management Act (PFMC SSC 2000). By the mid-1980s, many traditional groundfish stocks were being fished at maximum allowable levels and groundfish landings (with the exception of Pacific whiting) were in decline. In an attempt to protect individual stocks and meet the needs of seafood processors for year-round supply of fresh fish, the PFMC applied increasingly restrictive trip limits[2] to the trawl sector, a side effect of which was increasing discards of fish at sea. In addition, the entire fleet of trawl and fixed gear vessels was overcapitalized and faced the threat of further capacity increases from Alaskan vessels entering the fishery. At the urging of the groundfish industry to limit access to groundfish fishing, the PFMC began developing a program to restrict the number of participants in the fishery by issuing a limited number of fishing permits for groundfish, referred to as license limitation. After nine years in development by eight ad hoc and three standing committees, and after several challenges, a license limitation program was implemented in 1994. While the final program addressed many equity issues by lowering the minimum landings qualifications for license eligibility and including an "open access" component,[3] the net effect was to weaken controls over the overall size of the fishery (Hanna 1995a). Nevertheless, the program was the first step in creating limited property rights in the multispecies fishery through the definition of tiers of vested interests. It also defined who had rights to participate in the fishery, which provided the basis for subsequent property rights developments such as the trawl individual quota system currently under development.

Property Rights Impasse: Pacific Sablefish (*Anoplopoma fimbria*) **IQs**

The fixed gear sablefish sector (pots and longlines) of the groundfish license limitation program was not managed under trip limits as used in the multispecies trawl sector. Year-round landings were less important to this gear group, whose market was primarily export of frozen product. Within its quota allocation, the sablefish fishery was conducted as a derby competition until 1993, by which time the season had followed the predictable path of derby fisheries and had declined from six months to three weeks (PFMC 1994).[4]

In 1992 at the urging of the industry, the PFMC established a committee to develop an individual quota (IQ) program for sablefish and halibut and end the derby conditions. To provide security to IQ program participants, the fixed gear sector needed a guaranteed fixed proportion of the harvest quota for these species. The program development process was beset by problems, including high transaction costs of coordination and negotiation, incomplete representation of fixed gear interests, and a failure to address several equity concerns. Responding to these problems, the PFMC voted in 1994 to delay the program indefinitely. In 1996, the Sustainable Fisheries Act's moratorium on individual quota program development in federal fisheries prevented further consideration (Hanna 1995a). Subsequent actions taken by the PFMC amended the management plan to allow multiple vessel permits to be reassigned to a single vessel, a practice called permit stacking.[5] The trading of vessel permits approximated some of the properties of a tradable individual quota program in that it allowed more efficient scales of vessel operations and reduced fleet size (PFMC 2001).

Property Rights Retreat: Columbia River Fish Wheels

An older Pacific fishery illustrates retreat of property rights. Since the 1880s, the use of fish wheels, beach seines, and traps for commercial harvest of Columbia River salmon entailed the use of privately owned equipment and sites. Such methods were outlawed by Oregon in 1926 and by Washington in 1934. The prohibition took place through ballot measures in each state within a context of increasing salmon scarcity and rising transaction costs of fishing due to growing conflicts between fixed and mobile gear users. The legal actions had the effect of finalizing the reallocation of access to salmon runs from the private owners of fixed gears and exclusive fishing sites to mobile gear users who used vessels to catch salmon in open water using drift gill nets or troll gear. Long before the prohibitions of fixed gear took place, the property rights of fixed gear users were being attenuated by mobile gear users. In the late 1800s, mobile gear users were allowed to intercept salmon at will and there were no restrictions on their entry into the fishery, while fixed gear users bore the brunt of regulations (Higgs 1982, 65–68). The allowance of mobile gear entry represented a drawing back from property rights and a movement toward open access accompanied by technical prohibitions and a lowered productivity of capital. In the Depression-era political economic context, the final prohibition of fixed gear was an electoral triumph of "equity" (labor) over capital (Higgs 1982, 68).

Process over Structure

The structural characteristics of each of the three property rights systems, at least in their intent, reflected the four essential properties identified by Scott (1988) as exclusivity, permanence, transferability, and security. Two of the three programs, as designed, would score relatively high marks on Arnason's (2000) measure of quality (Q) of property rights with two notable exceptions. The groundfish license limitation program only weakly achieved effective exclusivity because of its emphasis on providing equitable access. Against precedent and owner expectations, the Columbia River fixed gear property rights failed in the attributes of title security and permanence, as in the end they were rendered moot by the Washington and Oregon electorate. The push for IQs in the sablefish fixed gear sector had three of the requisite property rights characteristics with the fourth, transferability, missing by design. The IQ program failed to move to implementation because of unresolved distributional concerns. The political inaction was followed within two years by a moratorium on developing new IQ programs contained in the 1996 Sustainable Fisheries Act.

The three cases illustrate that beyond structure, the processes of change by which systems of property rights are developed or maintained are supremely important.

Conditions for Change

The three examples of changing property rights in commercial fisheries illustrate seven critical conditions that influenced the direction of change:

- *Scarcity*: fish stocks reach a condition of scarcity relative to the demand for their use.
- *Transaction costs*: costs of allocation, operation, conflict reduction, or enforcement rise to a point of outweighing current benefits.
- *Affected interests*: industry, rather than managers, proposes the change in property rights.
- *User participation*: decision processes are fully representative, transparent, and based on consistent expectations.
- *Management support*: administrators recognize the potential for reductions in management costs under a new regime.
- *Enforcement*: a case can be made for more effective enforcement.
- *Equity*: notions of fairness among user groups are satisfied.

How far a fishery went along the path toward more fully specified property rights depended on the extent to which these conditions were met. The successful Pacific groundfish license limitation program met all seven conditions. The failed Pacific sablefish IQ program met five of the seven. The Columbia River fixed gear prohibition, which represented a retreat from property rights, met only three out of seven conditions.

All three fisheries were in a state of relative scarcity when the proposal for change was made, each fishery having experienced the "fishing down" process of virgin stocks under varying forms of derby competition. Under the derby competition, the benefits of rent accrued to the strongest competitors, those who caught the most fish the fastest, who then become the proponents of rights-based systems with allocations based on fishing history. In the case of Columbia River fixed gear, it was the greater political power of those excluded from the rent accrual process that in a later stage eventually prevailed. For all three, the time of the action was characterized by transaction costs of management, fishery operations, and conflict resolution that had risen to a level where some fishery interests found costs outweighing benefits.

In each fishery attempts to change property rights originated in the fishing industry, rather than with managers, although to varying degrees. The attempts were characterized by large-scale changes in ideas about open access: against open access, in the cases of groundfish and sablefish, and in favor of open access in the case of the fish wheels and other fixed gear. Proposals for change in the Pacific groundfish fishery, initiated by the trawl sector, were supported by most of the trawl and non-trawl fleet. With less unanimity, the sablefish IQ proposal was initiated by the large-scale producers with little support by small-scale operators. On the Columbia River, proposals for change were made not by the owners of fixed gear but by all other fishery interests who stood to gain by the reduction in harvest efficiency.

The groundfish and sablefish programs had the support of managers, who saw them as potentially cost-reducing through the resolution of allocation conflicts and the reduction of the size of the interest groups with which they had to interact. In contrast, Columbia River managers were faced with the cost-increasing potential of expanding the number of harvesters and harvest sites subject to monitoring and enforcement. However, managers were not the decision makers with regard to outlawing fish wheels. In both Oregon and Washington the public were the decision makers through their votes on state ballot measures.

Equity considerations played both a positive and negative role in the outcomes of the three cases. The groundfish license limitation program, as the first effort to define rights in West Coast fisheries, proceeded against a traditional path of open access. Participants in the program development process were sensitive to the political tensions caused by resistance to enclosure. They proceeded with extreme openness to equity concerns. As a result, the implemented program contained many equity attributes at the expense of efficiency. In contrast, the sablefish IQ program development process, in not ensuring full representation over all scales of operation, paid inadequate attention to equity concerns that were then expressed to decision makers in sufficient strength to stop the program. The Columbia River fixed gear prohibition is another example of the strength of equity concerns in fisheries. A ballot referendum in the context of the Great Depression found a public sympathetic to the notion of providing wider access to fishing opportunities even at the expense of efficiency and existing property rights.

Three Pacific Recreational Fisheries: Potential for Property Rights?

Three Pacific recreational fisheries illustrate allocation issues that would be amenable to solution through systems of property rights: Columbia River upriver spring Chinook, Oregon black rockfish and Klamath River fall Chinook.

Columbia River Upriver Spring Chinook

Upriver spring Chinook (*Oncorhynchus tshawytscha*) are large, bright fish with a high fat content, highly prized in both recreational and commercial fisheries. Harvest opportunities for spring Chinook are at a premium resulting in intense competition among commercial and recreational fishers (Hanna et al. 2006). Upper Columbia River spring-run Chinook are listed as "endangered" under the ESA. Harvest impacts on this run are managed through the allocation of allowable "impacts" to downriver commercial and recreational fisheries. An impact is an unintended mortality of incidentally caught and released fish from the upper Columbia River spring Chinook run. Impacts to listed fish are calculated as the percentage of the total listed population that represents mortalities as the result of fishing (ODFW/Washington Department of Fish and Wildlife [WDFW] 2005). Commercial salmon fisheries in the lower river need an upriver impact allocation to allow economically viable catch levels in target fisheries. Recreational salmon fisheries in the main stem of the Columbia River also need an upriver impact allocation to maximize their fishing opportunities when spring Chinook are most abundant (ODFW/WDFW 2005). The impact allocation is a type of bycatch quota, meaning that once the impact allocation is met fisheries in the lower river are shut down. The size of the impact allocation is proportional to the total upriver spring Chinook run size and ESA listed run size.

The non-Indian portion of the upriver catch was not formally allocated among commercial and recreational fisheries until 2002. The current arrangement of sharing impacts was adopted in 2004. The formula is based on the determination of preseason impacts to guide management of specific fisheries as well as in-season transfers of impact allocations between commercial and recreational fisheries as conditions in the fisheries change (ODFW/WDFW 2005). Within the total allowable impact, the states of Oregon and Washington have broad discretion to decide what allocation between recreational and commercial fisheries represents the public interest.

The recreational sector is concerned that too small a share of impact allocation will constrain its ability to maintain full seasons and continued growth. The commercial sector is concerned that the continuing expansion of the recreational sector will erode its access to the spring Chinook fishery and threaten its economic viability. Feelings on the issue are strong, and the absence of property rights to a share of the allowable impact in either sector results in a continuing negotiation over shares. These processes have high transaction costs for all involved.

At an intense public hearing in early 2006, the recreational sector requested the Oregon Fish and Wildlife Commission (OFWC) move away from the previous 60/40 recreational/commercial allocation split to provide the entire 2006 impact allocation to their sector. The OWFC spent over ten hours listening to public statements before deciding to allocate 55 percent of the allowable impact to recreational anglers and 45 percent to commercial fishers, with the ability to deviate 5 percent either direction to achieve to the extent possible the objectives of a full season for sports fishery and the maintenance or increase of the commercial catch (OFWC 2006).

Oregon Black Rockfish

Black rockfish (*Sebastes melanops*) is a long-lived, nearshore species of low productivity. It is listed as a "strategy species" in the Oregon Nearshore Strategy (ODFW 2006), indicating its identification as a nearshore species in greatest need of management attention. It lives in rocky sub-tidal habitat among diverse biological communities and is subject to both commercial and recreational harvest. Harvest limits are in place for both commercial and recreational sectors, and in-season adjustments in regulations are made to slow or stop harvest as these limits are approached (ODFW 2007). In addition, recreational and commercial harvest of black rockfish can be stopped by reaching a cap on associated species. Caps are expressed in poundage and are different for the recreational and commercial sectors. For example, as of August 30, 2006, black rockfish recreational harvest had reached 56 percent of the total cap, while recreational harvest of "other nearshore rockfish" was 68 percent of the total cap (ODFW 2006), making it likely that "other rockfish" limits will constrain recreational fishing for black rockfish.

Inadequate data on recreational and commercial harvest, particularly on species composition and geographic distribution of fishing over time, creates several uncertainties in the stock assessment of black rockfish (PFMC 2003). Commercial catch of black rockfish by the trawl, fixed gear, and open-access sectors is managed through trip limits adjusted within seasons to keep total (directed and incidental) catch within total limits. Over time, managers have reduced fishing in black rockfish areas by Oregon commercial fleets to provide additional nearshore access to recreational fleets. Oregon commercial vessels fishing in nearshore waters are now required to hold a limited entry permit with a nearshore fish endorsement (ODFW 2004). Washington has closed all state waters to commercial fishing for black rockfish (WDFW 1999).

Klamath River Fall Chinook

In August 2006, the Secretary of Commerce declared salmon fisheries from Oregon to California commercial fishery failures as a result of harvest restrictions put in place to protect returning Klamath River fall Chinook salmon (*Oncorhynchus tshawytscha*) (U.S. Department of Commerce 2006). Under emergency rule procedures the PFMC was able to allow commercial and recreational fishing at reduced levels. The management problem is the inability to differentiate weak Klamath River stocks from other healthier stocks when they are intercepted by ocean fisheries, leading to protective large-area closures that limit fishing on all stocks.

In response to the loss of commercial salmon fishing opportunities, representatives of Oregon State University and the Oregon salmon troll fleet developed a collaborative research project to assess the use of genetic markers to identify and avoid weak-stock bycatch in ocean fisheries. The project will also collect location and oceanographic data to understand relationships of physical and environmental factors such as ocean currents and stock location and movement. The new knowledge base of genetic, biological, and oceanographic information, allowing fine resolution spatial management in "real time," would also provide opportunities for new management approaches based on proportional shares of individual salmon runs. Recreational as well as commercial fisheries would be able to use, hold, or trade quota shares as mutually advantageous to other fishing sectors or areas (OSU COMES 2006). However, for a quota share system to receive political support from commercial and recreational sectors, both sectors would have to see such a program as being to their advantage. As yet, neither the commercial trollers nor the recreational anglers have expressed a collective interest in this approach.

Conditions for Change

As with the examples of changing property rights in commercial fisheries, the same conditions can be expected to influence whether a recreational fishery is likely to adopt more fully defined property rights.

Relative scarcity is a condition common to the salmon fisheries on Columbia River spring Chinook and Klamath River fall Chinook and to the groundfish fishery on black rockfish. Transaction costs in all three fisheries are increasing in the form of reduced fishing opportunities over space and time, greater gear restrictions, and rising conflicts with the commercial sector.

A key difference between the commercial and recreational examples is that the affected recreational interests have not yet expressed their discomfort with rising transaction costs in the form of an interest in property rights. There are a number of possible reasons for this omission. Recreational interests may see their competitive potential to acquire larger recreational allocations as being greater through the political process than through the obtainment of rights. In addition, the social learning about the economic benefits of property rights that Anderson (2000) has described in commercial fishery rights-based systems has not taken place. And finally, user group expectations may still be very much based on freedom of access that make discussion of rights-based systems more "out of step" than they are in commercial fisheries (Anderson 2000).

Pacific managers are not obviously aware or in support of the potential for rights-based systems in recreational fisheries as cost-reducing mechanisms. Equity notions of maximum access are evident in the way managers have chosen to allocate recent limited fishing opportunities. As Anderson (2000) noted, even in the Gulf of Mexico where recreational fisheries are of significant economic importance, managers had difficulty fitting the recreational red snapper fishery into a proposed individual transferable quota (ITQ) system. In addition, Pacific fishery managers are likely to identify inadequate systems for full catch accounting as a problem in developing and enforcing a rights-based system.

Overall, despite the restrictions on recreational fishing required by the conditions in the salmon and rockfish fisheries, only some of the conditions that determined change in the commercial fisheries are present. Without a more concrete demonstration of the cost-reducing or benefit-generating potential of property rights, it is unlikely that these recreational fisheries will follow a similar path in the near future. By allowing markets to take care of allocation, tradable property rights offer recreational fishery managers a potential means to remove themselves from the valuation/allocation anguish and offer recreational anglers the potential for more flexibility in fishing. Notwithstanding these potential benefits, neither managers nor anglers appear ready to initiate a rights-based approach.

Conclusion

This chapter has examined theoretical and empirical determining factors in the evolution of property rights in fisheries. Examples of fishery property rights development, blockage, and retreat have been described for commercial fisheries. In these processes a number of contextual conditions influenced the direction and pace of evolution as well as the eventual structural outcome.

These conditions are scarcity, transaction costs, the role of affected interests, user representation, management support, enforceability, and equity.

The evolution of rights-based fishery management approaches in recreational fisheries will be subject to the same influences. Three examples of Pacific recreational fisheries experiencing difficult allocation problems were described. These problems, although amenable to resolution through the application of property rights, have not yet led either anglers or managers to initiate this approach. Perhaps more importantly, the enabling conditions that motivated the development of property rights programs in commercial fisheries are only thinly in place. It is an open question as to when or how property rights will be used in Pacific recreational fisheries. Baseline awareness of the potential and possibilities of property rights systems in recreational fisheries may be the first educational task.

Notes

1. Recreational fishery interests are well represented on the PFMC, where five of eight non-agency seats on the council are currently held by recreational representatives (PFMC 2006a).

2. A trip limit defines allowable quantities of fish that may be landed by a vessel within a given period of time.

3. The open-access component of the groundfish fishery comprises vessels targeting groundfish without limited entry permits and vessels who target non-groundfish fisheries that incidentally catch groundfish. Trawl gear may not be used in the directed groundfish open-access fishery, but trawl gear may be used for target species such as pink shrimp, California halibut, ridgeback prawns, and sea cucumbers. The PFMC is now considering a license limitation program for the directed open-access fishery (http://www.pcouncil.org/facts/groundfish.pdf).

4. A derby fishery is one in which fishers competitively harvest the total allowable catch. Each fisher engages in a race for fish, trying to catch as much as possible as fast as possible, before the total quota is caught and the fishing season is closed. Derbies result in excessive capital investment, shortened fishing seasons, and unsafe fishing conditions.

5. Permit stacking allows non-trawl permit owners to register multiple limited entry fixed gear permits with sablefish endorsements to a single vessel. Sablefish permit stacking was designed to reduce capacity in the limited entry fixed gear sablefish fishery (http://www.pcouncil.org/control/controldates.html).

References

Anderson, Lee G. 1984. An Economic Analysis of Joint Recreational and Commercial Fisheries. In *Allocation of Fishery Resources*, ed. John H. Grover. Proceedings of the

Technical Consultation on Allocation of Fishery Resources, Vichy, France, April 20–23, 1980. Rome: U.N. Food and Agriculture Organization, 16–26.

———. 2000. Selection of a Property Rights Management System. In *Use of Property Rights in Fisheries Management*, ed. Ross Shotton. Proceedings of the FishRights99 Conference. Fremantle, Western Australia, November 11–19, 1999. *FAO Fisheries Technical Paper* 404/1 and 404/2: 26–38. Rome: Food and Agriculture Organization of the United Nations.

Arnason, Ragnar. 2000. Property Rights as a Means of Economic Organization. In *Use of Property Rights in Fisheries Management*, ed. Ross Shotton. FAO Fisheries Technical Paper 404/1, 14–25. Proceedings of the FishRights99 Conference Fremantle, Western Australia, November 11–19, 1999.

———. 2008. Harmonizing Recreational and Commercial Fisheries: An Integrated Rights-Based Approach. This volume.

Arrow, Kenneth J. 1974. *The Limits of Organization*. New York: W. W. Norton & Co.

Blamey, Russell. 2002. The Recreational Sector. In *Valuing Fisheries: An Economic Framework*, ed. Tor Hundloe. St. Lucia, Australia: University of Queensland Press, 113–65.

Criddle, Keith R. 2008. Examining the Interface between Commercial Fishing and Sportfishing: A Property Rights Perspective. This volume.

De Alessi, Michael. 1998. Fishing for Solutions. *IEA Studies on the Environment*, No.11. London: Institute of Economic Affairs.

Edwards, Steven F. 1990. An Economics Guide to Allocation of Fish Stocks between Commercial and Recreational Fisheries. NOAA Technical Report, NMFS 94, November.

———. 1991. A Critique of Three "Economics" Arguments Commonly Used to Influence Fishery Allocations. *North American Journal of Fisheries Management* 11: 121–30.

Eggertsson, Thrainn. 1990. *Economic Behavior and Institutions*. Cambridge: Cambridge University Press.

Gersick, Connie J. G. 1991. Revolutionary Change Theories: A Multilevel Exploration of the Punctuated Equilibrium Paradigm. *Academy of Management Review* 16: 10–36.

Hanna, Susan. 1995a. User Participation and Fishery Management Performance within the Pacific Fishery Management Council. *Ocean and Coastal Management* 28(1-3): 23–44.

———. 1995b. Efficiencies of User Participation in Natural Resource Management. In *Property Rights and the Environment: Social and Ecological Issues*, ed. Susan Hanna and Mohan Munasinghe. Washington, DC: World Bank, 59–67.

———. 2006. Implementing Effective Regional Ocean Governance: Perspectives from Economics. *Duke Environmental Law and Policy Forum* 16: 205–16.

Hanna, Susan, Gilbert Sylvia, Michael Harte, and Gail Achterman. 2006. Review of Economic Literature and Recommendations for Improving Economic Data and Analysis for Managing Columbia River Spring Chinook. A report to Oregon Department of Fish and Wildlife in fulfillment of ODFW Agreement No. 005-4132S-Wild. Institute for Natural Resources, Oregon State University, Corvallis, April.

Hannesson, Rögnvaldur. 2004. *The Privatization of the Oceans*. Cambridge, MA: MIT Press.
Higgs, Robert. 1982. Legally Induced Technical Regress in the Washington Salmon Fishery. *Research in Economic History* 7: 55–86.
Hirschman, Albert O. 1992. *Rival Views of Market Society and Other Recent Essays*. Cambridge, MA: Harvard University Press.
Hundloe, Tor, ed. 2002. *Valuing Fisheries*. St. Lucia, Australia: University of Queensland Press.
Iuddicello, Suzanne, Michael Weber, and Robert Wieland. 1999. *Fish, Markets and Fishermen: The Economics of Overfishing*. Washington, DC: Island Press.
Iverson, Edwin S. 1996. *Living Marine Resources: Their Utilization and Management*. New York: Chapman and Hall.
Larkin, Peter A. 1977. An Epitaph for the Concept of Maximum Sustainable Yield. *Transactions of the American Fisheries Society* 106(1): 1–11.
Leal, Donald R., ed. 2005. *Evolving Property Rights in Marine Fisheries*. Lanham, MD: Rowman & Littlefield.
National Marine Fisheries Service (NMFS). 2005a. *A Vision for Our Recreational Fisheries. NOAA Recreational Fisheries Strategic Plan 2005-2010*. Silver Spring, MD: U.S. Department of Commerce, National Oceanic and Atmospheric Administration, NMFS. (Available at http://www.nmfs.noaa.gov/ocs/documents/Fisheries_Strategic_Plan.pdf.)
———. 2005b. Fisheries of the United States 2004. *Current Fishery Statistics* No. 2004, Silver Spring, MD: U.S. Department of Commerce, National Oceanic and Atmospheric Administration, NMFS, Fisheries Statistics Division, Office of Science and Technology. (Available at http://www.st.nmfs.gov/st1/fus/fus04/index.html.)
North, Douglass C. 1990. *Institutions, Institutional Change and Economic Performance*. Cambridge: Cambridge University Press.
Oregon Department of Fish and Wildlife (ODFW). 2004. Fact Sheet: New Commercial Black Rockfish/Blue Rockfish Nearshore Fishery Limited Entry Permit (6/1/04). Marine Resources Program. (Available at: http://ftp.dfw.state.or.us/MRP/regulations/commercial_fishing/blackrf/blackblue_factsheet121003.pdf#search=%22Oregon%20commercial%20landings%20black%20rockfish%20%22.)
———. 2006. Oregon Sport Ocean Regulations for Salmon, Halibut and Other Marine Fish Species. (Available at http://www.nmfs.noaa.gov/ocs/documents/Fisheries_Strategic_Plan.pdf.)
———. 2007. Marine Regulations. Fish Division, Marine Resources Program. (Available at http://www.dfw.state.or.us/MRP/regulations/sport_fishing/ and at http://www.dfw.state.or.us/MRP/regulations/commercial_fishing/.)
Oregon Department of Fish and Wildlife (ODFW) and Washington Department of Fish and Wildlife (WDFW). 2005. Joint Staff Report Concerning Commercial Seasons for Spring Chinook, Steelhead, Sturgeon, Shad, Smelt, and Other Species and Miscellaneous Regulations for 2005. January 14.
Oregon Fish and Wildlife Commission (OFWC). 2006. Meeting Minutes, January 5–6. (Available at http://www.dfw.state.or.us/agency/commission/minutes/06/jan/010506%20Minutes.pdf.)

Oregon State University (OSU) Coastal Oregon Marine Experiment Station (COMES). 2006. A Proposal for Collaborative Research Conducted by the Oregon Salmon Commission and the Coastal Oregon Marine Experiment Station submitted to the Oregon Water Enhancement Board.

Pacific Fishery Management Council (PFMC). 1994. Draft Amendment 8 (Fixed Gear Sablefish Individual Quotas) to the Pacific Coast Groundfish Fishery Management Plan.

———. 2001. Permit Stacking, Season Extension, and Other Modifications to the Limited Entry Fixed Gear Sablefish Fishery. Amendment 14 to the Groundfish Fishery Management Plan. March. (Available at http://www.pcouncil.org/groundfish/gffmp/gfa14.html.)

———. 2003. Black Rockfish Star Panel Report. In *Status of the Pacific Coast Groundfish Fishery through 2003 and Stock Assessment and Fishery Evaluation*, Volume 1, August.

———. 2006a. Member Roster. (Available at www.pcouncil.org.)

———. 2006b. Pacific Council News. Summer 30(2). (Available at www.pcouncil.org.)

Pacific Fishery Management Council Scientific and Statistical Committee (PFMC SSC), 2000. Overcapitalization in the West Coast Groundfish Fishery: Background, Issues and Solutions. March 16. (Available at http://www.pcouncil.org/groundfish/gflibrary.html.)

Scott, Anthony D. 1988. Development of Property in the Fishery. *Marine Resource Economics* 5: 289–311.

Sutinen, Jon G., and Robert J. Johnston. 2008. Angling Management Organizations: Integrating the Recreational Sector into Fishery Management. This volume.

Tirole, Jean. 1995. *The Theory of Industrial Organization*. Cambridge, MA: MIT Press.

U.S. Department of Commerce. 2006. Commerce Department Declares Commercial Fishery Failure for Coastal Oregon and California. Press release, August 10. (Available at http://www.commerce.gov/opa/press/Secretary_Gutierrez/2006_Releases/August/10_KlamathDisaster_Release.htm.)

Washington Department of Fish and Wildlife (WDFW). 1999. Concise Explanatory Statement Regarding Live Fish Fishery Prohibition, WAC 220-20-010, December 15.

Williamson, Oliver E. 1985. *The Economic Institutions of Capitalism: Firms, Markets, Relational Contracting*. New York: Free Press.

———. 1994. Transaction Cost Economics and Organization Theory. In *The Handbook of Economic Sociology*, ed. Neil J. Smelser and Richard Swedberg. Princeton, NJ: Princeton University Press, 77–107.

———. 1996. *The Mechanisms of Governance*. New York: Oxford University Press.

Chapter 2

Recreational Fishing and New Zealand's Evolving Rights-Based System of Management

Basil M. H. Sharp

One of the leading challenges for New Zealand is the issue of incorporating recreational interests into its innovative fisheries management system. On the commercial side, over 90 species, or groups of species, are managed as 562 separate fish stocks under New Zealand's quota management system (QMS) with its emphasis on individual transferable quotas (ITQs). Under this rights-based system of management, now over twenty years old, commercial fisheries have come to play a prominent role in New Zealand's economy. In 2004, the seafood sector was New Zealand's fifth largest export earner when it posted NZ$1.2 billion in exports; generated NZ$150 million in domestic sales, contributed NZ$1.7 billion to gross domestic product; and, produced NZ$4.5 billion in total output (NZ Ministry of Fisheries 2006b). In addition, there are 2,500 seafood entities, providing direct employment for 10,500 full-time equivalent people. On the recreational side, surveys indicate that up to 20 percent of the population engages in marine recreational fishing annually. The main species pursued are snapper, blue cod, kahawai, rock lobster, paua, and scallop. Recreational fishing also contributes to the economy through business for equipment suppliers, charter boat operators, and tourist facilities.[1] Expenditures made by recreational fishers to catch five key recreational species have been estimated to be nearly NZ$1 billion per annum (South Australian Centre for Economic Studies 1999).

As the population concentration grows in areas such as Auckland there is increased pressure on the region's recreational resources, including coastal fish stocks. Notably, many of the marine species taken by recreational fishers are fished in competition with the commercial fishing sector. In a relatively small number of fisheries, such as the snapper and kahawai fishery off the northeast coast of the North Island, and the blue cod fishery at the top of the South Island, the recreational catch makes up a significant portion of the total annual catch.

Using policy instruments such as ITQs in commercial fisheries and recreational bag limits, the results of standard bioeconomic modeling have important implications for practical policy design.[2] The necessary conditions describing an optimal allocation of the flow of services from a given fish stock are described elsewhere in this book. For purposes here it is sufficient to state that at optimality the present value of net marginal benefits is equalized across the commercial and recreational sectors. In particular, at optimality the total harvest and the allocation of the harvest across the two sectors are determined simultaneously. Thus an allocation will fail the optimality test if the aggregate harvest is not set at the optimal level or the aggregate harvest is not allocated across competing interests such that the present value of net marginal benefits are equalized. The necessary conditions derived from bioeconomic modeling provide a framework for management of a shared fishery even though institutional structure is not explicit. Of course the main task or challenge for fishery management is to figure out whether or not an allocation mechanism can produce efficient outcomes and, in particular, whether the mechanism can capture and incorporate the necessary values attributable to commercial and noncommercial uses.

This chapter focuses directly on the crucial issue of mechanism design and reports on progress toward integrating recreational harvest into New Zealand's rights-based management regime (Hurwicz 1973). I begin with descriptions of New Zealand's QMS and developing recreational fishery policy. Three case studies are subsequently presented to highlight competition between the commercial and recreational sectors and the allocation process. Two of the case studies were litigated when the Minister of Fisheries attempted to adjust the level and intersectoral allocation of total allowable harvest. The third case study illustrates how cooperative management can work to satisfy the goals of both commercial and recreational interests. The chapter concludes with a brief overview of the challenge facing mechanism design.

Quota Management Framework

The QMS is a rights-based system of management with two key structural pillars: a total allowable catch (TAC) and ITQ rights.[3] Quota for fish stock included in the QMS is expressed in shares. At the beginning of each fishing season a fisher's ownership of ITQ shares is applied to the total allowable commercial catch (TACC) to determine the fisher's annual catch entitlement (ACE), expressed in tons. Both ITQ and ACE are tradable. Without going into detail, adjustments to the annual TACC affect the size of ACE. Thus increases (decreases) in the TACC translate into increased (decreased) holdings of ACE at the beginning of each fishing season.

When setting the TAC the Minister of Fisheries is required to provide for utilization and maintenance of the stock at or above maximum sustainable yield (MSY). At this stage of the statutory process the form of utilization is not specified but obviously includes commercial, recreational, and customary use. Competing utilization is, however, the focal point of the next step in the process. The annual TACC is set by subtracting catch allowances for customary and recreational interests and estimates of other sources of mortality caused by fishing (e.g., bycatch) from the TAC. For the purpose of analysis these noncommercial interests are summarized as the total allowable noncommercial catch (TANC). By law, the Minister must use the best available information when making the above decisions.

Let us now turn to two questions concerning optimality. First, does the QMS provide a basis for setting the optimal harvest? The answer is no. In economics it is well-known that MSY is not an optimal policy, because it is not based on maximizing economic rent from the fishery (Clark and Munro 1975). Second, does the QMS allocate the TAC in a way that balances the net present value of net marginal benefit across sectors? This question can be answered in two ways. If rights to utilization are transferable across competing interests, then we could rely on the market mechanism to determine who fishes and expect rights to gravitate to their most highly valued use. Legislation fails on this count because recreational fishers need no quota right to fish and thus there is no need to trade with commercial rights holders. Alternatively, if the best available information included estimates of the net present value of recreational fishing, then it would be possible to at least approximate the balancing required. To date, valuation of recreational fishing benefits is not carried out at the Ministry. Thus, there is little prospect of allocating the TAC in a way that maximizes net benefits.

Until the mid-1980s, legislation paid little attention to the rights of recreational fishers. Common law interpretations regarding the public right to fish for sport were based on *Malcolmson v. O'Dea* (1863). In that case, the House of Lords held that since the Magna Carta, the Crown could not establish exclusive fishing rights by grant and the public rights of fishery could not be overturned except by statute. Regulation of the public right to fish began in 1983 when legislation foreshadowed the 1986 implementation of the QMS. Current legislation exempts recreational fishers from requiring a fishing permit. Amateur fishing regulations set out the maximum number of fish that may be taken or possessed by any one person, in any day, and provides the basis for sanctions if bag limits and size restrictions are violated. The structure of property rights underpinning recreational harvest is quite different from that of commercial fishing rights. To commercially harvest fish, a fisher must have a fishing permit and ACE to cover the catch or pay the government a deemed value for the amount of harvested fish not covered by ACE. Recreational fishers do not require a fishing license, entry into the fishery is free, and their harvest is managed through bag limits and gear restrictions.

A flaw in the architecture of property rights emerges at the time an allowance is made for noncommercial interests. The value of a commercial fisher's property right depends on the quality of the right and, inter alia, the price of landed fish, harvesting costs, and stock abundance. Commercial fishers pay a levy to government that is designed to cover management and research costs associated with commercial fishing. Noncommercial fishers pay no license fee and entry is free. When these two interest groups compete in a shared fishery, setting the TACC is pivotal. Concatenating commercial and noncommercial interests at this stage, however, elevates the allocation process to a position where it cannot perform well without additional information, especially information on harvest and value. The Minister can with reasonable confidence, backed up with penalties associated with overfishing, monitoring information, and enforcement, expect commercial fishers to fish according to their entitlements and the rules governing harvest. Furthermore, quota markets provide information on the value of harvesting rights. In contrast, the recreational harvest faces no aggregate constraint. Even if recreational fishers conform to daily bag limits, population increases or growth in disposable income can easily result in the year's recreational harvest outstripping the annual recreational allowance. Moreover, information is lacking to anticipate impacts in a given fishery because the number of recreational fishers is not known, their harvest is not known, and the recreational value attached to the harvest is not revealed by the allocation mechanism. In short, information the Minister has when deciding the annual TAC "carve up" is not symmetrically distributed, and even if it were, it may not be commensurable because of the underlying property right regimes.

Recreational Fishing Policy

Prior to 1983, legislation provided little scope for active management of recreational fisheries. Management was aimed at preventing amateurs from taking excessive quantities of fish and shellfish and undersized specimens. The 1983 Fisheries Act, which foreshadowed but did not introduce the QMS, provided for a more integrated approach to fisheries management. Shortly after the QMS was implemented in 1986, the Ministry of Agriculture and Fisheries turned its attention to marine recreational fisheries (NZ Ministry of

Agriculture and Fisheries 1989). At the time, various forms of input controls were used to restrict amateurs, including size limits for retaining fish, daily bag limits, closed seasons, and gear restrictions.

The marine recreational fisheries policy announced in 1989 applied to all users of marine recreational fisheries and included both extractive and non-extractive users. The scope of the policy covered traditional forms of recreational fishing and diving as well as commercial big-game fishing and charter fishing. Firms providing commercial recreational opportunities were not included under the umbrella of QMS when introduced in 1986. Marine protected areas were to be established to cater for the needs of so-called non-extractive users. At the time, fishery management plans (FMPs) were seen as the primary fisheries management instrument, containing management objectives and controls for each fisheries management area. The planning process stalled for many years and was eventually abandoned.

The 1989 National Policy for Fisheries Management had two goals: (1) to conserve, protect, and enhance living marine and freshwater resources and the habitats on which they depend; and (2) to maximize the economic and social benefits from fisheries. Under this policy, the goal for marine recreational fisheries management was to maintain or improve marine recreational fisheries. Concern over the adverse effects of overfishing, presumably by commercial interests, on recreational fisheries was evident in the policy statement. Three principles underpinned the policy. The first principle stated that management must balance social, cultural, environmental, and economic costs and benefits arising from different uses. The second focused on limiting recreational harvest to within long-term sustainable yields. The third principle stated that the present rights of noncommercial fishers to fish in any waters will be maintained subject, of course, to input restrictions that might be necessary to maintain or improve fish stocks. Liberally interpreted, this principle underscored the public's belief that they had the right to freely enter and harvest seafood, subject to bag limits and restrictions on gear. The 1989 policy also clearly stated that the licensing of noncommercial users was considered inappropriate.

However, in discussing allocation, the 1989 policy stated that allocation of fishery resources should reflect the most beneficial use of the resource. What the drafters of the policy had in mind when using the term "most beneficial" is not clear, but this statement supported the view that a share of the sustainable harvest (TAC) would be allocated to recreational users. Furthermore, where a species is not sufficiently abundant, preference in allocating the catch was to be given to noncommercial fishing in areas readily accessible and popular with the public. Daily catch limits and other restrictions were to be used to achieve the allocation target within the recreational sector. Interestingly, spatial instruments were foreshadowed when the policy noted that areas could be set aside for recreational activity.

The New Zealand Recreational Fishing Council (NZRFC) is a nonprofit organization made up of delegates representing national and regional recreational fishing associations and clubs. Its primary activity is to advocate and represent the interests of any noncommercial marine fishers and to participate in the protection and scientific study of the aquatic environment, aquatic life, fish, and their habitats. These activities are financed from membership fees. In 1998, at the NZRFC annual conference, the Minister of Fisheries, a national member of Parliament challenged the NZRFC to work collaboratively with government to test the public's views about better defining recreational fishing rights and management responsibilities. The NZRFC accepted the challenge, and a joint NZRFC/Ministry of Fisheries working group (JWG) was formed to develop options to identify and secure New Zealand recreational fishing rights and responsibilities. After several rounds of stakeholder consultation and workshops, a draft public discussion paper was presented to the newly elected Minister of Fisheries, a member of the Labor Party. The Minister noted that licensing of recreational fishing was not Labor's choice of government policy but agreed to leave the option in the discussion document so as not to stifle debate. The Minister also noted that the NZRFC did not have an exclusive mandate to represent recreational fishing interests.

In July 2000, a public discussion document, entitled *Soundings*, was released by the working group. The purpose of *Soundings* was multifold: to more clearly specify the relationship between recreational, commercial, and customary entitlements; to encourage greater management responsibility by recreational fisher organizations; to develop better tools for spatial management; and to improve information on the recreational catch. Toward these ends, three managerial options were presented:

> Option 1: Discretionary share. This was the status quo option and provided for a recreational allowance made by the Minister, spatial management of recreational vis-à-vis commercial fishing through fishery management plans, and government management of recreational fishing.
> Option 2: Proportional share. This option stated that ongoing proportional shares in key recreational fisheries would be established and new coastal zones would provide preferential access for recreational fishers. Recreational fishing would continue to be managed by government.
> Option 3: Recreational management. Option 3 included Option 2 plus the establishment of recreational fishing management groups recognized by government, with the management of recreational fishers shared between government and recreational groups.

About 14,000 copies of *Soundings* were mailed out by the JWG. However, an informal grouping of recreational fishers with like-minded interests formed and promoted their views under the banner of Option 4. The form letter distributed by the Option 4 group attracted a total of 61,178 submissions offering varying levels of support. Option 4 supported no licensing of recreational fishers, recreational priority right over commercial fishers for free access to a reasonable daily bag limit in legislation, the ability to exclude commercial fishers from recreationally important areas, and the ability to devise plans to ensure future generations enjoy the same or better quality of rights while preventing fish conserved for such purposes being given to the commercial sector.

The Option 4 group saw the fundamental conflict as one between recreational fishers and commercial fishers. They saw no conflict with other noncommercial fishers such as customary fishers. Their position was centered on a recreational entitlement being an inalienable right. Once this entitlement was recognized in law, the group advocated a national, elected body to represent recreational interests and to provide management of the recreational harvest. Because the entitlement was viewed as a "public good," the group argued that central government should be the primary source of funding, supplemented by the groups themselves. Compulsory funding mechanisms were rejected.

In general, the commercial fishing industry supported a change to better defined recreational rights but not at the expense of established commercial rights and a proportional share of rights for key fisheries. Some industry groups advocated a proportional share with trading between commercial and recreational interests possible. Industry opposed preferential recreational rights and spatial management of areas that excluded commercial fishing. Many industry submissions argued that the recreational sector should share the cost of management research and compliance, as industry does.

This outcome left the JWG with little to move forward with in terms of mutually supported goals by commercial and recreational interests. Their report to the Minister, in March 2001, included recommendations to further develop shared fishery policy, to better define public share and access to fisheries, to not include any form of licensing in future policy development, and to explore ways of better measuring the recreational harvest.

Late in 2006, the Ministry of Fisheries launched work on a public discussion paper on shared fisheries (NZ Ministry of Fisheries 2006a). The latest discussion paper acknowledges that shared fisheries are under pressure and that effective management is undermined by poor information on recreational harvest and its value. Although the language is couched in noneconomic terms, one of the key ideas advanced in the discussion paper is that shared fisheries should be managed in a way that produces best value. If interpreted to mean total economic value, then the conditions of optimality outlined above should hold. Currently, information on the recreational catch is obtained using phone-and-diary surveys, which have produced estimates that can vary up to 300 percent. Proposals to improve information include more surveys to capture catch and participation data, requiring charter operators to report their harvest through log books and nonmarket valuation surveys. In order to account for the greater value amateur fishers place on size of fish and abundance, it was suggested that the biomass in shared fisheries might be managed at levels above that consistent with MSY. The discussion paper reassures amateur fishers that their basic right to catch will be retained and suggests that a minimum tonnage could be retained and given a priority right over commercial fishing. If adopted this would result in a minimum allocation for recreational fishers, provided, of course, that the TAC was tracking toward MSY as required in law; the balance, if any, would be allocated as TACC available to commercial fishers. Setting and adjusting initial entitlements was considered difficult because of information gaps. Priority shares in favor of recreational interests and fixed share adjustments were considered unlikely to maximize value. Options considered include resetting allocations after valuation studies and following a negotiation process. Although there are already tools available for spatial management, the paper advances ideas for noncommercial coastal fishing zones and fisheries plans for shared fisheries developed by both commercial and noncommercial sectors. The creation of an amateur fishing trust was advanced as a means of strengthening the voice of amateur fishers. The policy document gives no clear indication as to the structure and financing of the trust. An amateur fishing trust could, for example, be formed around the NZRFC to be recognized as a voice of recreational fishers when it comes to their position on annual allowances for recreational harvest and ongoing policy as it relates to recreational management, and research. Activities of the trust could be financed from membership fees combined with government subsidy. At the time of this writing, no firm proposals for managing shared fisheries have emerged.

Tale of Three Species

Incorporating the interests of commercial and noncommercial fishers in shared fisheries is a vexatious issue facing fisheries policy makers. Beginning in 1983, numerous proposals have been floated for public discussion. Well over twenty years later, New Zealand has what is considered by many an ineffective recreational fishing policy. Of the estimated 400,000 marine recreational harvesters, two-thirds live in the more populated northern region of New Zealand, a region close to fisheries management area one. As noted earlier, the Minister of Fisheries is required to make an allowance for utilization. Two court cases highlight the difficulty of using the TAC mechanism as the primary instrument for making catch allowances between the commercial and recreational sectors. Importantly, these two cases will show that institutional design generates asymmetric information on values. The market for tradable rights within the QMS provides information on the value of commercial harvest, but no mechanism exists to provide value estimates of the recreational harvest. Litigation highlights the governance vacuum created by the lack of a robust policy to deal with recreational interests in shared fisheries. An enhanced scallop fishery provides an illustration of the potential benefits of management that integrates both commercial and noncommercial interests.

Snapper

Snapper (labeled SNA in the QMS) is an inshore species with high commercial value. About 85 percent of the commercial harvest is exported, returning around NZ$30 million in export earnings. Snapper is also highly sought after by recreational fishers and an important source of food for Maori. The northeastern management area for snapper (SNA 1) adjoins New Zealand's largest metropolitan area with around 1.3 million residents, and population projections see the region growing to 1.7 million by 2026. Not surprisingly, the region is home to New Zealand's largest population of recreational fishers. With snapper keenly sought after by both commercial and recreational fishers, this fishery was to become the first battleground.

Snapper was among the group of species first introduced into the QMS in 1986. The TACC was set at 4,710 tons, which was 55 percent of the established catch history to allow for stock rebuilding. Subsequent decisions by the Quota Appeal Authority saw the TACC increase to 6,010 tons in 1991. The TACC was exceeded by over 500 tons in the 1993 fishing year. Some of this was the result of quota holders carrying forward up to 10 percent of catch entitlements not caught from previous years, as permitted by legislation. There is a reasonably well-developed market for quota rights. The ratio of annual lease-to-sale prices through the 1990s shows a trend toward the opportunity cost of capital measured by 90-day bank bills, which indicates that traders in the quota market are cognizant, at least to some degree, of what they can earn elsewhere in the economy (Batstone and Sharp 2003).

Although the first recreational bag limit was set in 1983, the recreational fishery was left to itself with minimal monitoring and enforcement. In 1985, a national recreational bag limit of thirty snapper per person per day was set with a minimum legal size (MLS) for fish retention of 25 centimeters. Limited evidence exists on recreational catch rates. During the early 1990s, the average catch per recreational fisher per hour varied from 0.4 to 0.6 snapper. In 1993, the individual daily bag limit was reduced to twenty snapper, followed by a further reduction in 1995 to fifteen with the MLS set at 27 centimeters. A 1994 survey by the NZRFC found that the average catch per day for each person was 1.9.

Table 2.1 illustrates the projected trend in the biomass for one of the sub-stocks in SNA 1. Biological modeling suggests that this sub-stock is recovering and now exceeds the biomass of 65,000 tons, which is considered necessary to support MSY. Table 2.1 also shows the TACC being reduced once over the period, with commercial fishers landing close to 100 percent of the TACC each year. The TANC, set to cover recreational and Maori customary use, has not changed, although bag limits have been reduced and MLS increased over the period. Estimates of recreational harvest, based on telephone interviews of fishers keeping diaries are shown in table 2.1. The enormous increase in estimated recreational harvest from 1997–1998 to 2000–2001 is, unfortunately, attributed to survey design deficiencies. Even if the noncommercial harvest stays within its allowance, 34 percent of the TAC is allocated to noncommercial interests.

Table 2.1
Stock Biomass, Allowances, and Catches for Snapper in SNA1

Year	Stock Estimate	TACC	Commercial Catch	% Caught	TANC	Recreational Catch
1994–95	40,000	4,938	4,831	98	2,600	2,857
1995–96*	40,000	4,938	4,941	100	2,600	—
1996–97*	42,000	4,938	5,049	102	2,600	—
1997–98	45,000	4,500	4,524	100	2,600	2,324
2000–01	50,000	4,500	4,347	97	2,600	6,242
2001–02	55,000	4,500	4,372	97	2,600	6,738
2002–03	57,000	4,500	4,484	99	2,600	—
2003–04	60,000	4,500	4,466	99	2,600	—
2004–05	65,000	4,500	4,637	103	2,600	—

Notes: Unless indicated otherwise, all data are in tons. The TANC includes allowance for customary (300) and recreational use (2,300). The stock biomass estimates are indicative of projections for one sub-stock within SNA1 and should not be interpreted as an accurate record of biomass estimates for SNA1. The estimated B_{MSY} for this particular sub-stock is 65,000 tons.
* Minister's proposal to reduce TACC to 3,000 tons, leaving TANC at 2,600 tons.
Source: New Zealand Fishing Industry Association (Inc) and Others v. The Chief Executive Officer of the Ministry of Fisheries and Others (1997).

In 1995, a Ministry of Fisheries Advice paper for TACC and management controls for the 1995–1996 fishing year looked specifically at reducing the recreational harvest. It was estimated that a reduction in the bag limit to ten fish per day would give a 4 to 6 percent reduction in recreational take (about 130 tons), and a reduction to six fish per day would give an 8 to 12 percent reduction in recreational take (about 260 tons) (*New Zealand Federation of Commercial Fishermen (Inc) and Others v. The Chief Executive Officer of the Ministry of Fisheries and Others* 1997). The Ministry noted that recreational catch was not controllable by bag limits. At the time of setting the TAC and its allocation in late 1995, the Minister noted that the low recreational catch per unit effort (CPUE) was not satisfactory.

Rights-based fisheries management provides a clear focus of the issues involved and the commercial consequences of adjustments to harvest levels. Nine days before the beginning of the 1995–1996 fishing season, the Minister announced the following allocations: the TAC was to be 5,600 tons, the TACC was reduced from 4,938 tons to 3,000 tons, and the customary Maori and recreational harvests were set at 300 tons and 2,300 tons respectively. Commercial fishing interests sued and were successful in preventing implementation. The issue did not go away, however, and five days before the beginning of the 1996–1997 fishing season, the Minister announced an identical decision. Again, the court prevented implementation. This time the case was considered in the High Court and later in the Court of Appeal.

Details on the legal debate that followed the 1995 decision are worth reviewing. The above allocation proposed by the Minister was first contested in the High Court. Applicants for a judicial review of the Minister's decision included commercial fishing interests and Maori. Respondents included the Minister of Fisheries, the Chief Executive Officer of the Ministry of Fisheries, and the New Zealand Recreational Fishing Council. Commercial interests claimed, inter alia:

1. A reallocation from ITQ holders to noncommercial fishers, without compensation, was improper.
2. Sustainable yield at the current stock level is 92 percent of MSY and the 8 percent increase in yield at the Minister's proposed MSY is theoretical. In practice yield at that MSY would not be detectably different.
3. The commercial sector was penalized for absence of constraint on recreational fishing.
4. Legislation did not offer priority to recreational harvest when it came to allocating the TAC.
5. The 1996 Fisheries Act placed a duty on administrators to control the noncommercial catch within the allowance made.

6. The economic impacts and the benefits of reduction were not adequately considered.

As to be expected, Maori involved in commercial fishing supported industry claims. Furthermore, Maori noted that the ITQ they held were part of a compensation package provided for under the Deed of Settlement Act (1992) and that the Crown had a duty to protect the value of the settlement (Sharp 1997). Maori argued that the value of the quota reduction (about NZ$14.6 million) was a transfer from Maori to recreational fishers. In addition, Maori protested lack of priority given to customary take.

According to the Ministry of Fisheries (2005) the SNA1 court decision was made under the 1983 Fisheries Act and confirmed that the Minister has wide powers of discretion in deciding the allocation to each sector. In particular, the Minister has the power to determine the nature and extent of priority between recreational and commercial interests and to vary the proportion of the commercial harvest relative to the recreational harvest once an initial allocation has been made. The court findings highlighted a significant relationship between the rights of commercial fishers and the power of the Minister. Action by the latter could impact the value of rights held by the former. The 1996 Fisheries Act requires the Minister to quantify and allow for noncommercial interests, including recreational interests, when setting the TACC. In late 1996, a Ministerial briefing document commented on the problem and a possible solution, as follows: "How to more explicitly recognize recreational rights is an issue needing comprehensive policy analysis . . . a greater number of recreational sector representatives see some benefit in a better definition of recreational entitlements" (*New Zealand Federation of Commercial Fishermen (Inc) and Others v. The Chief Executive Officer of the Ministry of Fisheries and Others* 1997, 57).

Did the Minister give priority to the noncommercial catch and, if so, was that an erroneous interpretation of the law? The judge was not persuaded that the Minister operated on an assumption of law that noncommercial has priority ahead of commercial interests. Furthermore, the judge was not persuaded that the Minister gave priority, in effect, by failing to impose effective controls that would or could control recreational effort. In April of 1997, all claims by commercial interests were dismissed with judgment in favor of the Minister. In the meantime, however, implementation of the Minister's decision had to wait on progress in the courts, the effect of which was to leave the TACC at 4,938 tons through fishing year 1997–1998. The recreational daily bag limit of nine and minimum legal size of 27 centimeters was left in place.

Litigation then moved on to the Court of Appeal, where commercial interests lined up against the same group of respondents. The Court of Appeal turned its attention to an earlier requirement for the Minister to consider the possibility of the Crown acquiring quota and not exercising the right to harvest, thus reducing the need to lower the TACC as much and dampening the adverse impact on industry. The Court of Appeal found that the Minister had been wrongly advised in this regard and that both TACC-setting decisions were flawed. Notably, express reference in the 1996 Fisheries Act to the Crown retaining or obtaining rights, and not using the rights, was repealed in 1996, after both TACC-setting decisions.

Perhaps of greater interest is the approach of the court to allocating the TAC. First, the Court of Appeal noted the Minister's duty to move the stock toward a level that can produce MSY. Furthermore, "when deciding upon the time frame and the ways to achieve that statutory objective the Minister must consider relevant social, cultural and economic factors" (*New Zealand Fishing Industry Association (Inc) and Others v. The Chief Executive Officer of the Ministry of Fisheries and Others* 1997, 14). Second, the Court also noted that the legislation is structured in a way that requires the Minister to first set the TAC and then make allowances for other interests before fixing the TACC. In particular, the court saw no reason why the ratio of commercial to noncommercial harvests could not be altered. Retreating from a position of strict proportionality, the court held that the Minister must act reasonably to prevent the savings resulting from a TACC reduction being lost to recreational fishing. The Court of Appeal set aside the Minister's 1995 and 1996 decisions. This result further reinforces the earlier point made about mechanism design for allocations between sectors. The QMS, as structured and implemented, did not perform well in shared fisheries.

The snapper case provides legal precedent to dealing with competing claims to the TAC. The economic implications of the court's finding include (1) when setting the TAC, decision makers should assess the costs and benefits when analyzing the alternative trajectories that will take the stock toward MSY; (2) there is no a priori reason why the proportion of commercial harvest to recreational harvest cannot change and relative value matters; and (3) if the harvest of one sector is reduced for conservation reasons, then the ability of the other sector to free ride on the benefits of this decision should be curtailed.

Kahawai

Access to kahawai stocks was to become the second battleground. Kahawai are a schooling pelagic species occurring mainly in coastal areas. Known for its fighting ability when hooked, kahawai is considered an excellent game fish. The commercial industry utilizes kahawai in combination with skip jack tuna to sustain its purse seining operations. It is also harvested using trawling and long lining, both as target and bycatch species. According to an affidavit from V. Wilkinson (*The New Zealand Recreational Fishing Council and Others v. The Minister of Fisheries and Others* 2007, 6), kahawai is a popular eating fish in the domestic market. This view, however, was not supported in an affidavit from R. Winstanely (*The New Zealand Recreational Fishing Council and Others v. The Minister of Fisheries and Others* 2007, 12), who reported that kahawai are not considered a quality eating fish but a sport fish and are often used as bait in lobster pots.

Before 1990, competitive catch limits for kahawai were not set and the commercial fleet landed between 4,300 and 9,600 tons per season. Commercial catch limits were imposed for the fishing season 1990–1991, and these limits endured through August 2004. Recreational daily bag limits and controls on customary fishing were also in place. Until 2004, commercial harvesters of kahawai were managed under a permit system. In 2004, upon introduction into the QMS, the Minister of Fisheries was required to (1) announce a kahawai TAC for each quota management area, (2) make allowances for noncommercial interests (TANC), and (3) set a TACC. The Ministry of Fisheries' preferred policy position was to use catch history and to reduce commercial and recreational utilization in equal proportions if any reduction was necessary for taking the stocks toward MSY. Setting the TAC was central to the introduction of kahawai into the QMS. Uncertainty over recreational utilization coupled with uncertainty over the status of kahawai stocks led the Minister to adopt a cautious approach.

In August 2004, prior to the start of the 2004–2005 fishing season, the Minister announced a 15 percent reduction in both the commercial and recreational fishing levels for kahawai. At that time, it was suggested that to achieve the reduction in harvest by recreational fishers a reduction in their daily bag limit might be necessary. However, this did not happen and the recreational daily bag limit remained at twenty per person per day. At the time, it was noted that very little information existed on the magnitude of recreational harvest. In 2005, the Minister announced that the annual TAC was to be reduced by 10 percent to allow stocks to rebuild. Again, the reduction in catch was shared among all sectors. The Minister did not change recreational bag limits and noted that there was no evidence that the recreational sector was catching the allowance assigned to it.

Table 2.2 shows the TACs for each QMS, the noncommercial allowance, and allowance for other sources of mortality. Like snapper, the utilization of kahawai is greatest in management area one (KAH 1). In KAH 1, the recreational allowance exceeded the TACC and accounted for 40 percent of the TAC. Over all fisheries management areas, the TACC was 43 percent of the TAC and the recreational share 36 percent. Recreational fishers were limited to a daily bag limit of twenty fish per day.

Table 2.2
Proposed Allowances for Introducing Kahawai into QMS, 2004

	KAH 1	KAH 2	KAH 3	KAH 4	KAH 8	KAH 10	Total
TACC	1,480	710	490	10	635	10	3,335
Recreational	1,580	510	300	5	380	5	2,780
Customary	790	255	150	3	190	3	1,391
Mortality	60	35	20	0	5	0	120
TAC	3,910	1,510	960	18	1,210	18	7,626

Note: All data are in tons.
Source: The New Zealand Recreational Fishing Council and Others v. The Minister of Fisheries and Others (2007).

In 2004, a legal challenge initiated by the New Zealand Big Game Fishing Council and the New Zealand Recreational Fishing Council sought a declaratory judgment from the High Court as to what the law is regarding the Minister's powers and decisions under the 1996 Fisheries Act. In particular, the claimants alleged:

1. Current use estimates were not based on the best possible information when making the initial allocation.
2. The allocation failed to consider the social, cultural, and economic factors relevant to recreational fishers.
3. The decision wrongfully relied on quantitative estimates (i.e., tons) of recreational use to infer qualitative interest viz. the abundance and quality of sport fish.

Commercial fishers counterclaimed as follows:

1. The Crown has not used the tools available to it to ensure sustainable utilization, particularly with respect to managing the recreational harvest.
2. There were errors in advice given to the Minister regarding the status of the kahawai stocks.
3. Restrictions applied to the commercial sector with no effective management of the recreational sector are an indirect allocation to the recreational sector.
4. Compensation may be sought for the economic loss suffered by commercial fishers.

Three aspects of managing a shared fishery are highlighted by the above claims: (1) information, (2) recreational fisheries management, and (3) the balancing of commercial and recreational use. First, effective fisheries management relies on robust information on stock abundance and sectoral harvest levels. Although information on stock abundance is common to both commercial and recreational fisheries, information on recreational harvest was of lower quality. Information asymmetries make it even more difficult to set initial entitlements. The kahawai case centers on setting initial entitlements when introducing a species into a rights-based system of management. Each side sought a greater share of the TAC through the legal system. Second, commercial harvest is effectively constrained within the QMS. In contrast, while the harvest of individual recreational fishers is limited by daily bag limits, there is no limit on the aggregate recreational harvest. Third, improving the efficiency of management highlights the need to balance economic values associated with use. It is relatively straightforward to estimate the economic value of commercial use and, indeed, the QMS as an allocation mechanism generates this information. Not so with recreational use. Clearly, aggregate recreational harvest does not measure value. It is important to note that this particular case centered only on whether or not the Minister acted in accordance with his statutory powers and obligations. The court did not have jurisdiction over whether the Minister's decisions were correct or not. Table 2.3 provides a summary of recreational and commercial claims as they relate to these three aspects.

When setting the TACs for 2004 and 2005, the judge was not satisfied that the Minister had failed to have due regard to social, cultural, and economic factors considered relevant so as to result in the stock being moved to a level that could support MSY. Turning to the setting of the TACCs, the judge relied

Table 2.3
Summary of Claims

	Recreational Interests	Commercial Interests
Information	Over-reliance on current use	Flawed estimates of recreation use
	Undue weight to commercial interests	No cost-benefit analysis
Management	Common law right to fish	Ineffective management
	Can overturn public right by statute	Obligation to control recreation use
Allocation	Proportion of TAC not fixed	Implicit reallocation to recreation
	Regulatory takings not compensable	Regulatory taking compensable

on evidence showing that kahawai is of low value to the commercial sector and, by implication, that consumer demand for kahawai is low. The judge noted that the economic valuation exercise, which set the commercial value of a right alongside an estimated nonmarket value associated with recreational harvest, showed that the value of kahawai for recreational fishers was comparatively eleven times greater than its value for commercial interests. However, the value imbalance did not find favor with officials who claimed that catch history is a more certain method for allocation. The judge held that a policy preference for catch history cannot take precedence over a mandatory requirement to adopt a utilization approach. Reliance on catch history led the Minister to place undue weight, when setting the TACCs, on the potential effect of catch reductions on commercial operations, which was not considered to be the correct statutory test.

Thus the Minister's decisions in 2004 and 2005 were unlawful to the extent that the Minister (a) fixed the TACCs for kahawai without having due regard to the economic, social, and cultural well-being of people; and (b) failed, without giving any or proper reasons, to consider advice from the Ministry of Fisheries to review bag catch limits for recreational fishers. The court directed the Minister to reconsider or review his 2005 decisions to take into account the terms of the declaration of unlawfulness.

The kahawai case study illustrates the challenge of establishing initial entitlements and providing acceptable allowances for recreational fishing in a shared fishery. Establishing initial entitlements in unshared fisheries is reasonably straightforward, provided, of course, that one accepts the position that initial entitlements should be based on historical catch. However, in the shared kahawai fishery, recreational interests considered that their initial share did not reflect the relative value of their interest and that the Minister did not adequately account for their interest when setting the TACCs.

Enhanced Scallop Fishery

Like many valuable New Zealand fisheries the northern South Island scallop fishery was managed as an open-access regime. As economic theory would predict, vessel numbers increased and harvest levels eventually decreased. Attempts at controlling aggregate harvest included issuing a limited number of vessel licenses and, on occasion, closure to commercial fishing (Arbuckle and Drummond 2000). Although the fishery remained under restrictive licensing when the QMS was introduced in 1986, successful seeding trials pointed to potential gains that could come from industry cooperation in enhancing the scallop stock. In his review of the QMS,

Table 2.4
Scallop Allowances, 2007

Fishery	TAC	TACC	Recreation	Daily Bag Limit
SCA1	75	40	8	20
SCA7	827	747	40	50
SCA8	48	22	8	20

Note: Daily bag limits are shown as legal-sized individual scallops; other data are tons of scallops.
Source: NZ Ministry of Fisheries (2007).

Peter Pearse (1991) identified the fishery as a prime candidate for involving fishers in management of the resource. In 1992, the Challenger scallop fishery was introduced, in modified form, into the QMS. Self-imposed levies on scallopers collected by the Ministry were held in trust to finance implementation of the plan.

In the mid-1990s, the Challenger Scallop Enhancement Company (CSEC) signed an agreement with the Ministry of Fisheries to implement an enhancement plan approved by the Minister. Scallop quota owners formed the company with the aim of augmenting the stock, managing the timing of within season harvest, and harvesting at a level that approximates maximum economic yield instead of maximum sustainable yield. The CSEC established and supported a recreational advisory group that participated in annual harvesting decisions. Actions by the CSEC to enhance opportunities for recreational harvest include closing sub-areas of the fishery to commercial use over the summer vacation period. These areas are available to recreational scallopers as part of a cooperative effort to share the fishery. Activities of the company are underpinned by self-financed research, compliance, and management, subject to approval by the Minister of Fisheries (Sharp 2005). Enhancement activities by the CSEC provide a beacon for the management of shared fisheries.

Table 2.4 shows the allowances made for three scallop management areas, with only SCA7 being enhanced. The enhanced scallop fishery is by far the most productive, reflecting the gains from enhancement and the natural advantage the area has in terms of the marine environment. With a TACC of 747 tons, it has been estimated that the southern scallop fishery can produce up to 60 percent of New Zealand's scallop harvest when operating at full capacity. The area also offers recreational divers the highest tonnage of scallops and the highest daily bag limit. Recreational harvest is open from mid-July through mid-February. Scallop enhancement is considered a forerunner of other shared fisheries such as rock lobster, snapper, and salmon (Booth and Cox 2003).

Policy in Shared Fisheries

In shared fisheries the Minister must make an explicit allocation of the TAC to both commercial and noncommercial interests. By law, ITQ (commercial) rights are attenuated because they are share rights to the TACC. If the Minister lawfully reduces the TACC, then the property embodied in the share right has not been invaded. In other words, the economic value of the share right is bounded by law and the legal actions of the regulator. In the snapper case, the Court of Appeal offered the following advice to the Minister: "All we wish to say for the future is that the Minister would be wise to undertake a careful cost/benefit analysis of a reasonable range of options available to him in moving towards MSY" (*New Zealand Fishing Industry Association (Inc) and Others v. The Chief Executive Officer of the Ministry of Fisheries and Others* 1997, 23). A similar message was conveyed in the kahawai case: "This is an exercise in judgment, to be carried out by weighing up and balancing the recreational fishers' right to provide for their social, economic and cultural well-being by fishing for kahawai against the extent, if any, to which the peoples, in the sense of the wider general public, well-being is served by commercial interests in satisfying consumer demand" (*The New Zealand Recreational Fishing Council and Others v. The Minister of Fisheries and Others* 2007, 30). The message is clear: adjustments to the TAC, including reallocations between sectors, should be guided by cost-benefit analysis.

These two court cases present the management agency with a problem because there is little information on relative values in recreational fishing to guide the Minister's decision. Quota markets generate information on the commercial value of harvesting rights, but there is no independent source of information on the value attached to the recreational harvest. This is a fundamental problem with the current institutional structure. Information on the relative net benefits of quota allocation across the sectors is lacking. Moreover, incentive compatibility is asymmetric, and controllability, both in a practical and technical sense, exists only on the side of commercial fishing and is, therefore, one-sided.

To some degree, recreational fishing is rival and the benefits are excludable. The notion of an exclusive club which shares impure public goods and requires membership restrictions owing to potential overexploitation is one interesting approach for achieving greater institutional compatibility between recreational and commercial interests (Olsen 1965, Buchanan 1965). In their review of the economic theory of clubs, Sandler and Tschirhart (1980) define a club around the notion of a voluntary group deriving mutual benefit from sharing production costs, members' characteristics, and/or excludable benefits. Well-defined recreational fishing rights, especially the right to exclude nonmembers, would be necessary for a voluntary group to form. Clubs endowed with an allocation of rights and the ability to exclude nonmembers would be a way of closing the QMS with respect to recreational fishing.

The idea of applying the economic theory of clubs to recreational fishing is not new and perhaps was the motivation behind the Ministry's idea of establishing an Amateur Fishing Trust (NZ Ministry of Fisheries 2006a). To date, two variants have been suggested. Pearse (1991) proposed use of existing Fish and Game legislation. Regional Fish and Game councils are Crown Entities established under the Conservation Act 1987. They report to the Minister of Conservation. Councils represent the interests of fresh water anglers and hunters and provide coordination of the management, enforcement, enhancement, and maintenance of sports fish and game. Activities of Fish and Game Councils are funded from the sale of angling and hunting licenses. Freshwater angling is controlled by license, anglers pay a license fee, and daily bag limits apply. However, the existing Fish and Game Council structure was not favored for two principal reasons: (1) legislation limits Councils to freshwater fisheries, and (2) their institutional structure is not consistent with sound recreational fisheries management.

Sutinen (1996) suggested formation of recreational quota management clubs to manage the recreational harvest component of the noncommercial allocation. Commercial fishers have formed quota owner associations (e.g., the Challenger Scallop Enhancement Company), and the formation of clubs to represent recreational fishers and cater to recreational demand seems a plausible solution to the existing dilemma. Although it is legal for a group (possibly a nonprofit group) to form, buy quota, and harvest fish under the QMS, there are few incentives for recreational fishing clubs to voluntarily form because there is no license fee payable for the right to harvest marine species. The research and management necessary to determine the recreational allowance is supplied at no cost to the recreational fisher, and the cost of measuring, monitoring, and enforcing recreational activities is also provided by the taxpayer. To date, the total allocation for recreational fishing has increased over time. What are the incentives for clubs to form?

Progress toward a resolution rests on a number of key questions that policy makers must address. First, can a centralized mechanism approximate an optimal allocation—or even an allocation that satisfies both recreational and commercial interests? Second, how should initial entitlements be established for each sector? Third, what is the best adjustment process considering the inevitability that the TAC will change in the future? Fourth, what is the appropriate shape and architecture of recreational fishing rights? Finally, how should the supply of recreational entitlements, arising out of stock research and stock management activities, be financed? A robust information generating mechanism on harvest and values is essential. Sutinen's proposal could possibly achieve this through internal information gathering on members as well as through quota trading between clubs and between sectors, but there is no political will for establishing amateur fishing clubs with quota ownership. In the absence of licensing, the problem of estimating recreational harvest from sample surveys introduces uncertainty into the data. Surveys would have to be reasonably frequent, adding to the expense of management.

Picking up on the Challenger company scallop fishery example, a club of sorts that represented both recreational and commercial interests might offer a solution. Even if the club's quota holdings were subject to collective approval, members would be confronted with the costs of acquisition and management. A club might choose to hold a portfolio of quota rights, including both shared fishery quota rights and deepwater quota rights. Quota decision making would internalize commercial and recreational values, and decisions could be aimed at maximizing asset value. As a design mechanism, the club could provide dynamic guidance on recreational value relative to competing claims on the TAC. Establishing the initial entitlement could be based on historical harvest in a way similar to when ITQs were first introduced in 1986. This, of course, would shift the problem of allocation from the Minister of Fisheries to the club, thereby confronting members, commercial and recreational alike, with the opportunity costs associated with different allocations of the TAC. However, to reinforce earlier points, a club would have to be able to exercise its right to exclude nonmembers. In the absence of excludability, the recreational fishery would become an open-access fishery and economic value would be dissipated. Furthermore, if recreational and commercial interests are not joined and coordinated somehow, conservation gains in one sector would become an opportunity for free riding in the other.

Conclusion

The primary function of rights-based governance is to guide the decisions of fishers toward the allocation of a sustainable harvest surplus that is to some degree efficient. In an unshared fishery full economic efficiency is beyond the reach of New Zealand's QMS. Nevertheless, in unshared fisheries the QMS provides a basis for harvesting rights to be exercised by relatively more efficient fishers. Mechanism design within purely commercial fisheries is relatively straightforward. Even if a regulator sets an allowable harvest, as required in the case of New Zealand's QMS, the mechanism will generate useful information on relative profitability and state of the stocks. Who gets the right to fish is determined endogenously.

To operate efficiently, rights-based governance must be closed with respect to competing claims on the harvestable stock. If rights are not differentiated then transfers across sectors will see rights move toward relatively higher valued uses. This, of course, removes the need to separately specify allowances for commercial and recreational fishing. Monitoring and compliance would continue to be a significant activity. On the other hand, if rights are differentiated and not tradable across sectors then the allowance must be made through the TACC-setting mechanism. Making net-benefit maximizing allowances for both commercial and recreational fishing depends a great deal on the information available to the Minister. In case of commercial fishing, the QMS generates information on harvest and value. However, in the absence of recreational licenses and records of landings, the Minister has little, if any, information on which to base a decision. Furthermore, with unlimited entry and a harvest constrained only by bag limits and gear restrictions, aggregate harvest from recreational fishing is not capped, unlike the commercial sector which is output constrained. The different structure of rights underpinning the commercial and recreational sectors does not concatenate to solve the problem and places an undue burden on the TACC-setting mechanism. With sufficient pressure, competing recreational claims can dissipate economic surplus in the fishery.

Recent litigation in New Zealand well illustrates the difficulties a government agency has when it comes to setting an allowance for noncommercial interests. Currently, there is no support for licensing or transferable rights-based approaches for recreational fishing in New Zealand. While the range of organizational structures that might be used to effect closure is large, indications from the recreational sector suggest a partial solution might be found in spatial closures to commercial fishing, an approach not favored by commercial fishers.

While predicting future developments in recreational policy is hazardous, there are signs that commercial charter operators might be required to hold ITQs or at least report their harvest. Licensing may become a reality provided the issue of recognizing the interests of recreational fishers, beyond simple measures of catch, is resolved. If licensing were to come about, then managers would, at the very least, know the population of recreational fishers. This in turn would better facilitate survey design to estimate harvest and, if necessary, recreation values. However, cooperative management, along the lines of the CSEC model, would require collaboration between sectors underpinned by legislation that more clearly defines recreational fishing rights and empowers management to exclude nonmembers. Given that recreational fishers currently enjoy open access at no cost, there is little prospect of formal systems of cooperative governance, such as clubs, in the immediate future.

Notes

1. A large number of charter fishing vessels operate in areas such as Auckland's Hauraki Gulf and the Marlborough Sounds. These are included in the recreational fishing category because they do not sell fish, but rather supply services to the recreational market.

2. There is a reasonably well-developed literature on the economics of recreational fishing. Anderson (1993) provides a comprehensive analysis of the utilization and management of recreational fisheries. Bishop and Samples (1980) and McConnell and Sutinen (1979) illustrate the theoretical aspects of conflicts between recreational and commercial interests. Their work is particularly important because it provides an economic framework for allocating the optimal harvestable surplus across the two sectors. Sutinen (1993) builds upon this work by providing the necessary conditions for an optimal allocation of fish when enforcement is costly.

3. Only the key components of the QMS are described here; a more detailed description is contained in Sharp (2005).

References

Anderson, Lee G. 1993. Toward a Complete Economic Theory of the Utilization and Management of Recreational Fisheries. *Journal of Environmental Economics and Management* 24: 272–95.

Arbuckle, Michael, and Kim Drummond. 2000. Evolution of Self-Governance within a Harvesting System Governed by Individual Transferable Quota. In *Use of Property Rights in Fisheries Management*, ed. Ross Shotton. Proceedings of the FishRights99 Conference. Fremantle, Western Australia, November 11–19, 1999. *FAO Fisheries Technical Paper* 404/1 and 404/2: 370–82. Rome: Food and Agriculture Organization of the United Nations.

Batstone, Chris J., and Basil M. H. Sharp. 2003. Minimum Information Management Systems and ITQ Fisheries Management. *Journal of Environmental Economics and Management* 45: 492–504.

Bishop, Richard C., and Karl C. Samples. 1980. Sport and Commercial Fishing Conflicts: A Theoretical Analysis. *Journal of Environmental Economics and Management* 7: 220–33.

Booth, John D., and Owen Cox. 2003. Marine Fisheries Enhancement in New Zealand: Our Perspective. *New Zealand Journal of Marine and Freshwater Research* 37: 673–90.

Buchanan, James M. 1965. An Economic Theory of Clubs. *Economica* 32(125): 1, 14.

Clark, Colin W., and Gordon R. Munro. 1975. The Economics of Fishing and Modern Capital Theory: A Simplified Approach. *Journal of Environmental Economics and Management* 2: 92–106.

Hurwicz, Leonid. 1973. The Design of Mechanisms for Resource Allocation. *American Economic Review* 63: 1–30.

McConnell, Kenneth E., and Jon G. Sutinen. 1979. Bioeconomic Models of Marine Recreational Fishery Harvests. *Journal of Environmental Economics and Management* 6: 127–39.

NZ Ministry of Agriculture and Fisheries. 1989. *National Policy for Marine Recreational Fisheries*. Wellington, June 1989, 12.

NZ Ministry of Fisheries. 2005. *Shared Fisheries Policy Document*, Wellington, December 2005, 13.

———. 2006a. *Shared Fisheries: Proposals for Managing New Zealand's Shared Fisheries: A Public Discussion Paper*. Ministry of Fisheries, Wellington, 24.

———. 2006b. *Snapper (SNA)*. (Available at http://services.fish.govt.nz /fishresourcespublic/Plenary2006/SNA_06.pdf.)

———. 2007. *Scallop (SCA)*. (Available at http://www.fish.govt.nz/en-nz/SOF/Species .htm?code=SCA&list=name.)

Olsen, Mancur. 1965. *The Logic of Collective Action: Public Goods and the Theory of Groups*. Cambridge, MA: Harvard University Press.

Pearse, Peter H. 1991. *Building on Progress: Fisheries Policy Development in New Zealand*. Report prepared for the Minister of Fisheries, Wellington, NZ.

Sandler, Todd, and John T. Tschirhart. 1980. The Economic Theory of Clubs: An Evaluative Survey. *Journal of Economic Literature* 18(4): 1481–521.

Sharp, Basil M. H. 1997. From Regulated Access to Transferable Harvesting Rights: Lessons from New Zealand. *Marine Policy* 21(6): 501–17.

———. 2005. ITQs and Beyond in New Zealand Fisheries. In *Evolving Property Rights in Marine Fisheries*, ed. Donald R. Leal. Lanham, MD: Rowman & Littlefield, 193–211.

South Australian Centre for Economic Studies. 1999. *The Value of New Zealand Recreational Fishing*. Adelaide and Flinders Universities, South Australia, November, 114.

Sutinen, Jon G. 1993. Recreational and Commercial Fisheries Allocation with Costly Enforcement. *American Journal of Agricultural Economics* 75(5): 1183–87.

———. 1996. *Recreational Entitlements: Integrating Recreational Fisheries into New Zealand's Quota Management System*. A Report to the Ministry of Fisheries, Wellington, NZ.

Cases Cited

Malcolmson v. O'Dea, 11 ALL ER 1155 (NZ House of Lords 1863).

New Zealand Federation of Commercial Fishermen (Inc) and Others v. The Chief Executive Officer of the Ministry of Fisheries and Others, CP237/95 and CP294/96 (NZ High Court 1997).

New Zealand Fishing Industry Association (Inc) and Others v. The Chief Executive Officer of the Ministry of Fisheries and Others, CA82/97 and CA83/97 (NZ Court of Appeal 1997).

The New Zealand Recreational Fishing Council and Others v. The Minister of Fisheries and Others, CIV 2005-404-4495 (NZ High Court 2007).

Chapter 3

Can Transferable Rights Work in Recreational Fisheries?

Hwa Nyeon Kim, Richard T. Woodward, and Wade L. Griffin

Although innovations in the management of commercial fishing are being introduced around the world, recreational fisheries are still either unmanaged or managed using a set of policy tools with known problems. The use of these tools might be due to presumptions that more effective management of recreational fisheries would be too difficult to achieve in practice or perhaps that recreational fisheries are not worthy of careful consideration because their impact is not consequential. We now know that the second of these assumptions is not valid for a number of marine recreational fisheries in the United States (Coleman et al. 2004).

In recognition of the importance of recreational impacts on selected species, managers are beginning to apply tighter restrictions on the recreational sector. As shown in table 3.1, managers in the red snapper fishery in the Gulf of Mexico have decreased the daily bag limits, increased the minimum size limits for retaining fish, and increased the length of seasonal closures to limit recreational harvests. Even with the tighter restrictions, however, the total estimated recreational harvests still regularly exceed the annual quota allocated to the recreational sector. Hence, it seems that regulations are likely to become more restrictive and, as this occurs, anglers will become increasingly dissatisfied with management of the fishery.

Based on research supported in part by the National Marine Fisheries Service and by the Texas Agricultural Experiment Station.

Table 3.1
Changes in Recreational Red Snapper Regulations

Year	Size Limit (total length in inches)	Daily Bag Limit (no. of fish)	Season Length (days)	Recreational Allocation / Quota (million pounds)	Recreational Harvest (million pounds)
1991	13	7	365	1.96	1.94
1992	13	7	365	1.96	3.03
1993	13	7	365	2.94	5.29
1994	14	7	365	2.94	4.26
1995	15	5	365	2.94	3.25
1996	15	5	365	4.47	3.57
1997	15	5	330	4.47	5.41
1998	15	4	272	4.47	5.76
1999	15	4	240	4.47	5.51
2000	16	4	194	4.47	3.92
2001	16	4	194	4.47	4.52
2002	16	4	194	4.47	5.32
2003	16	4	194	4.47	4.58
2004	16	4	194	4.47	5.08
2005	16	4	194	4.47	4.59

Sources: Hood and Steele (2004, 16); data for 2004 and 2005 provided by Vivian Matter, NOAA NMFS Fishery Reporting Specialist, August 15, 2006.

In this chapter, we will consider one promising alternative to current recreational fisheries management—a system of transferable (fishing) rights (TRs). We begin with an overview of some of the problems of the suite of traditional management tools and then discuss in general terms the advantages of applying a TR approach to recreational fisheries management. Next, we discuss the experience that we have found documented in the use of TRs in fishing, hunting, and pollution control programs, highlighting some important lessons. Although relatively simple in principle, the success of TRs in recreational fisheries depends on a number of potentially critical institutional details. We provide a discussion of nine critical questions that must be addressed in the design of such a program, followed by a proposal for applying a TR program to the red snapper recreational fishery in the Gulf of Mexico.

The Management of Recreational Fisheries

For as long as economists have studied environmental and natural resource issues, they have criticized the use of strict command-and-control approaches to managing various resource commons. Essentially the critique is that a policy that holds all participants to a single, inflexible method of control does not allow for heterogeneous parties to find ways to achieve policy goals at lower costs or yielding greater benefits. Traditional approaches to management of recreational fisheries fall into this category.

Three main policy instruments are used to manage recreational fisheries: seasonal closures, minimum size limits for retaining fish, and daily bag limits. There are two primary motivations for a seasonal closure. First, it is sometimes justified based on the biological cycle of the species. For example, harvesting during the spawning season might be prohibited. In this case, a closure might be economically efficient as the harvest of a single pregnant fish during spawning can have large impacts on future stocks. Seasonal closures are also used to reduce total harvest. In this case, their use is economically inefficient. The inefficiency arises because inevitably some anglers will prefer to move their effort from the open season to the closed season, yet doing so would not impact fishing mortality. The loss of this opportunity for anglers with no impact on mortality lowers benefits and is therefore inefficient.

Referring to table 3.1 again, the length of the Gulf of Mexico red snapper recreational fishing season was reduced by over a third between 1997 and 2000 in an effort to reduce fishing effort. There is evidence that this move resulted in an inefficient temporal reallocation of effort. Figure 3.1 presents the distribution of fishermen surveyed in six two-month periods in 1997. We cannot infer from these data the distribution of angler trips during the year, but it does show that Gulf fishers do fish year round.[1] The figure also shows the dates the red snapper fishery has been closed in recent years. The current closure dates, between November 1 and April 20, correspond to times when a proportion of anglers used to like to fish but could no longer do so—a lost benefit.

The other two management tools used widely in recreational fisheries are size and bag limits. Virtually all managed fisheries have one or both of these regulations in place. The motivation behind these tools is multifaceted. Bag limits are used as a means for a more equitable distribution of the catch among anglers. They also create an obligatory guideline for what most would consider "good sportsmanship" in limiting one's catch on a given day. Placing a minimum size limit for retaining fish can help protect younger fish that survive catch and release, allowing for increased stock recruitment. As with bag limits, the size limit can coincide with what is generally considered good sportsmanship. Size and bag limits can, therefore, be the right tools for the objectives being pursued.

On the other hand, when they are used as a means to reduce overall fishing mortality, they are not necessarily the right tools. Panels A and B in

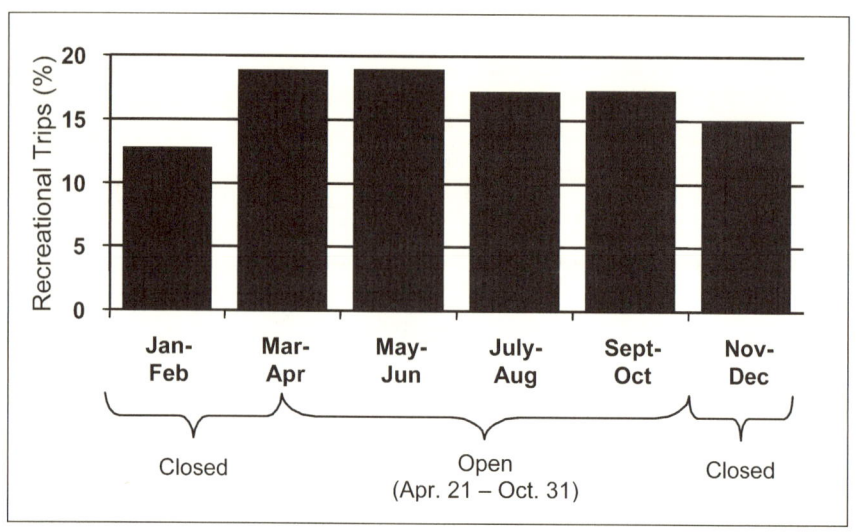

FIGURE 3.1
Percentage of Recreational Trips, 1997
Source: NMFS (1997).

figure 3.2 present the results from Woodward and Griffin's (2003) simulation analysis obtained from the General Bioeconomic Fisheries Simulation Model (GBFSM). GBFSM is a multi-species, multi-region simulation model in which growth and catch rates vary by both size and age of the fish. Because of this flexibility, the model can estimate the consequences of different combinations of bag- and size-limit policies. The panels show the impacts of different bag- and size-limit policies in terms of the present value of recreational consumer surplus on the horizontal axis and the stock in year 20 of the simulation. In panel A, release mortality rates are assumed to be quite low. In this case, size and bag limits are essentially substitute policies; that is, increasing the size limit has an effect similar to decreasing the bag limit in that both are used to increase future fish stocks at a cost to the immediate welfare of anglers in the fishery. In panel B, the release mortality rates are assumed to be much higher. In this case, increasing the size limit can actually be counterproductive because it leads anglers to discard more fish. When a significant share of discarded fish die, a policy of increasing the size limit does not necessarily lead to a reduction in overall fishing mortality.[2]

The analysis' results displayed in figure 3.2 are based on the assumption that anglers respond to bag limits by fishing until their bag limit is reached and then stop fishing. Under this assumption it follows that bag limits do not cause anglers to discard fish. In reality, however, bag limits can lead to discards of smaller fish if anglers hope to fill their bag with a larger fish. Hence, like

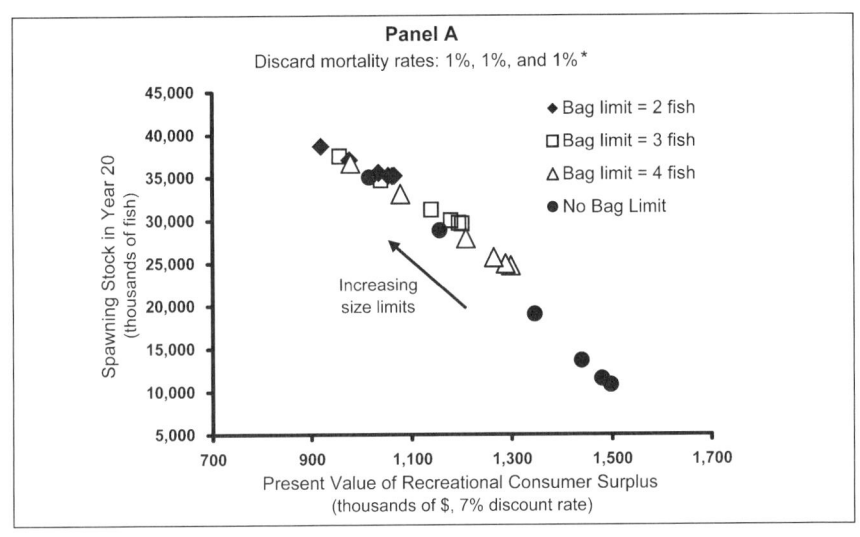

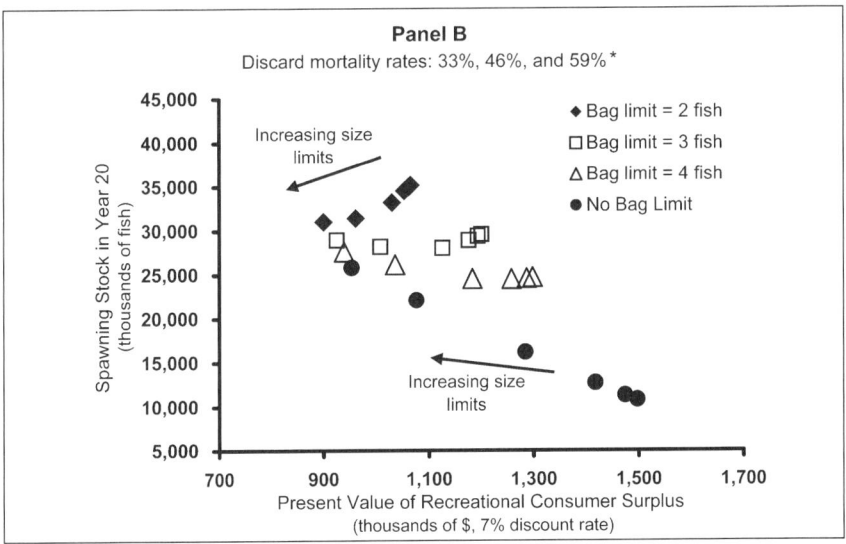

FIGURE 3.2
Simulated Consumer Surplus and Spawning Stocks, Year 20 (under alternative policy options)
*Mortality rates are for fish caught at three depths: 0–5 fathoms, 6–10 fathoms, and more than 10 fathoms.
Source: Richard T. Woodward and Wade L. Griffin, "Size and Bag Limits in Recreational Fisheries: Theoretical and Empirical Analysis," *Marine Resource Economics* 18, no. 3 (2003): 239–62. Reprinted with permission.

size limits, when discard mortality is significant, bag limits too are likely to be inefficient policies, because opportunities to catch fish in future periods are reduced from a failure to control fish mortality in the present. In the presence of release mortality, any policy that encourages discarding will be handicapped in its ability to achieve stock recovery goals.

The Potential for Transferable (Fishing) Rights

Among the policy alternatives for controlling use of public resources, a system of transferable rights (TRs) has received the most attention in the United States and, more recently, throughout the world. From air pollution to commercial fisheries, TRs are increasingly being promoted as the first option to consider when new goals are proclaimed in cleaning up the commons. The basic principle in any TR program is that a limited number of rights to use a public resource (such as air quality or a fishery) is made available. These rights can then be traded in a market so that in the end the rights are held by those who value them most. As with any market, one based on TRs will function properly if the property rights are comprehensively assigned, exclusive, transferable, and secure (Hanley, Shogren, and White 1997).

The first to propose a system of transferable rights to address environmental problems were Thomas Crocker (1966) and John H. Dales (1968). They recognized the potential of a system of TRs in meeting policy goals more effectively than command-and-control approaches. In addition, they pointed out that implementing such a scheme would require deliberate actions by government to create the program. Of particular importance in creating a successful TR program is the definition of what Dales (1968, 797) calls the "asset unit," defined as "the smallest physical amount of the asset to which it is practicable to apply property rights." Here, we interpret the asset unit somewhat more broadly to include the full set of rights and responsibilities that are embodied in the TR.

In some situations, defining the asset unit is relatively straightforward. When applied in commercial fisheries in a standard individual fishing quota (IFQ) program,[3] the asset unit is defined in terms of quantity of fish for each rights holder based on the total allowable catch (TAC) set for the year and the percentage share of the TAC held by the rights holder. In the national sulfur dioxide trading program, credits coincide with pounds of SO_2 emissions. The appropriate asset unit is not so straightforward in the case of recreational fisheries, where there are a number of questions concerning the characteristics of the asset unit that must be addressed if a TR program is to be established.

Before addressing these questions, it is useful to look at various applications of TR programs in pollution control, fishing, and hunting. These applications give insights into how TR programs for recreational fisheries might be designed under a variety of circumstances.

Transferable Rights in Pollution Control

The most prominent example of a TR program in pollution control is the U.S. SO_2 trading program, established under Title IV of the 1990 Clean Air Act. Since the second phase of the program began in 2000, more than 2,000 electric generating plants have been involved in the program, and a liquid market in the rights to emit SO_2 has resulted. A critical part of the success of the SO_2 program has been the availability of high quality monitoring through a continuous emission monitoring system that records emissions on an hourly basis. The monitoring program has made it possible to define the asset unit in the program in terms of measured SO_2 emissions. The large scale of the market and the high level of accuracy on actual emissions have made it possible to have a fluid TR market with very low transaction costs (Pérez Henríquez 2004).

The experience with TRs to control water quality in many ways has been a complete counter-example to the SO_2 program. The Lake Dillon program, which was created in 1984 to control phosphorous loading into Lake Dillon Reservoir in Colorado, provides an illustration (Woodward 2003). Two main sources of phosphorus were involved in the program: discharge from four publicly owned treatment works (POTW) and nonpoint discharge from septic systems throughout the region. A restriction prohibits POTWs from selling their surplus rights to emit phosphorous, so only nonpoint sources can generate credits for sale. Water quality credits are created when an underground septic system is eliminated by connecting the home to one of the POTWs, which are able to eliminate a greater share of the phosphorus before it reaches the lake. Since it is impossible to measure the exact phosphorus load from each septic system and there is an absence of records on historic loadings, credits cannot be given for actual phosphorus reductions. Instead, the asset unit in this program is based on practices: a credit is granted for each home connected to a POTW system. This approach ignores the high degree of variability relating to the quality of the original septic system, the water flow in the household, geographic proximity to the lake, and other factors. It is well known that quantifying credits based on practices rather than actual loadings leads to economic inefficiencies (e.g., Ribaudo, Horan, and Smith 1999). Nonetheless, because of the difficulty of monitoring nonpoint loadings, practice-based credit systems are used in all existing water quality trading programs that involve nonpoint sources. Water quality trading programs have also been challenged by the fact that the size of their markets is often quite small because they are naturally constrained to a single watershed. In the Lake Dillon program the first real transaction did not take place until fifteen years after the program was introduced. Most other water quality trading programs have been similarly limited (Woodward, Kaiser, and Wicks 2002).

Although perhaps far removed from fisheries, these two pollution examples provide interesting insights that can lead to (relative) success or failure in a TR program. The successful SO_2 program has a large market with an asset unit tied directly to the pollution load to be regulated. In contrast, water quality trading programs have small numbers of traders and asset units that are tied to practices rather than loads. As we consider the application of TRs to recreational fisheries, we must remember that all TR programs do not work as smoothly as the SO_2 program.

Transferable Rights in Commercial Fisheries

In commercial fisheries, TR instruments that have grown in use in recent years are individual fishing quotas (IFQs). Scott (1988) identifies four ways that IFQs can be an improvement over existing regulations. First, administrators need not concern themselves as much with dictating to fishers what gear type to use, but can instead focus on the issue of concern: the long-run management of the stock. Second, a quota system removes the incentive among fishers for a "race to the fish" that is so common in the traditionally regulated fishery. Third, a quota system can be preferred in the management of mixed stocks by allowing specialization in some species while at the same time purchasing rights to catch non-targeted fish. Finally, in an IFQ program, fishers, as quota holders, have an interest in supporting an effective monitoring and enforcement program.

Four nations use IFQs extensively in their commercial fisheries. These are Iceland, New Zealand, Canada, and Australia (e.g., Shotton 2001; Leal 2002). Other nations such as the United States, Greenland, and the Netherlands use them for some species. Along with Iceland, New Zealand is considered a pioneer in the use of IFQs. Batstone and Sharp (1999) provide a thorough review of the New Zealand's quota management system, which covered initially 26 species in 1986. The system has since expanded to include 93 species and 550 stocks for both inshore and deepwater regions (Leal, De Alessi, and Baker 2006). Initially, IFQs were designated as a fixed tonnage of fish per year for each quota holder based on historical harvests, but they were converted to percentages of the TAC for each fishery in 1990. Batstone and Sharp (1999) find evidence that the programs have been quite successful in adding economic value to fisheries, but they also point out that there are numerous practical challenges to implementing an IFQ: from recognizing historical rights of the Maori, the indigenous people of New Zealand, to managing fisheries outside the IFQ system.

IFQs are not, however, always preferred by fishers in certain fisheries suffering under a traditional approach to management. Rossiter and Stead (2003) discuss the problems in the demersal fishery in northeast Scotland. The fishery is closed each year once the TAC has been reached by fishers' aggregate catches, which has resulted in the fishery being closed earlier each successive year. Landings are falling, leading to economic decline, falling employment in the industry, and lower income levels. Based on interviews with fishers, the authors conclude that difficulties can be attributed largely to the race to the fish that now occurs as fishers try to catch as many fish as possible before the fishery is closed. It would seem that this fishery could benefit from an IFQ system, but the fishers do not favor such an approach. Instead, they favor regulations that restrict the number of days at sea for each fisher in a season, primarily because they believe that it would be easier to police. Rossiter and Stead and the fishers they interviewed recognized that a day-based approach would introduce some inefficiencies—for example, fishers facing such an approach may fish longer hours during their allowed days at sea and raise fishing costs by investing in more powerful vessels to maximize their catch. But the fishers preferred such a method because of its transparency and the flexibility that is allowed under such an approach. Time will tell whether these perceptions hold up in practice and whether days-at-sea regulations prove effective in controlling catches.

In addition, Clark, Munro, and Sumaila (2007) have shown that if fishers anticipate the introduction of an IFQ program in their fishery at some time in the future, all of the economic advantages of IFQs can be dissipated by excessive entry into the program in the pre-IFQ period. More generally, any time there is a perception that a TR program may be introduced, potential stakeholders will have incentives to alter their behavior in order to secure standing in the market that is to be created.

Transferable Rights in Recreational Fisheries

The National Research Council (1999, 212) recommended that attention should be given to the implications of recreational participation in fisheries, and to consider the potential application of IFQs in recreational fisheries. The literature and experience with any TRs, let alone IFQs, in recreational fisheries is quite limited, however. Sharp (1998) provided some initial ideas about how a recreational TR program might be structured and addressed some of the practical issues regarding allocation and monitoring. Sutinen, Johnston, and Shaw (2002) and Sutinen and Johnston (this volume) provide a more in-depth discussion on structure and issues and draw on relevant experiences from both the commercial and recreational sectors.

The recreational TR program that deserves the most attention is the proposed charter IFQ program for Alaska's halibut fishery, which Sutinen, Johnston, and Shaw (2002, 9) describe as "the sole U.S. template for the design of joint commercial-recreational rights-based management." Our review of this program offers insights into design elements, program merits, and practical problems that led to its demise. Criddle (this volume) identifies regional recreational catch allocations of the program,[4] which are some of the characteristics of the TR program's asset unit.

Other noteworthy characteristics of the program's asset unit include the following:

- The measuring unit of recreational IFQ is the number of fish, in keeping with current regulations.
- The charter IFQs are to be issued to charter owners (or to people who leased a vessel from an owner) who carried clients in 1998 or 1999 and 2000.
- The two-fish daily bag limit and the two-day possession limit of four fish per charter angler are retained.
- Charter quota shares are not to be sold to the commercial sector.
- Commercial individual fishing quotas can be traded to the charter sector with the added stipulation that commercial IFQ pounds be translated to recreational numbers of fish based on average weight.
- The program is not to affect non-charter halibut anglers.

As with any TR program, there were many practical issues that needed to be addressed, and these are evident in the minutes of the North Pacific Fishery Management Council's (NPFMC) committees (NPFMC 2001b, 2003). For example, enforcement issues such as prior notice of landings, the time window for offloading a vessel's landings, and what reports should be required of processors were considered (NPFMC 2001a). Resolution of these issues essentially define other characteristics of the asset unit by establishing the responsibilities associated with a TR in the program and, through monitoring, establishing the program's integrity. Such characteristics also affect the size of the program's transaction costs that would arise.

Initially passed by the NPFMC in 2001, the charter IFQ program was rescinded in 2005. Criddle (this volume) attributes the demise of the program to issues associated with the initial allocation. Despite the rule, noted above, that allocation would be based on effort from 1998 or 1999 and 2000; the exclusion of new entrants was controversial and, in the end, led to the program's cancellation. This experience in Alaska should not be forgotten as efforts are made to use TRs in other recreational fisheries.

Transferable Rights in Hunting

Although there is no example yet of IFQ-type TRs in practice for recreational fishing, there are various examples of less sophisticated TRs for recreational hunting and fishing. Johnston et al. (this volume) survey many of these applications. At the high end of the scale are programs in California and Colorado which allow market transfers of game tags for species such as deer and elk. Private landowners in these "ranching-for-wildlife" programs are issued a set number of game tags, depending on quality of habitat and game populations. They then sell, at market prices, the tags to hunters wishing to hunt on their lands. In Texas, the game tags are issued directly to hunters by the state for token fees, but there is essentially a free market for access rights to hunt the game on private lands. In this case, the sellers are the landowners and the buyers are the hunters willing to pay market prices for access rights (Leal and Grewell 1999).

The Kansas nonresident hunting market, which has many similarities to TR programs for recreational fishing identified in Johnston et al. (this volume),[5] is another example. This program, which began in 2000, has resulted in a competitive market in recreational hunting (Taylor and Marsh 2003). Characteristics of the asset unit in the Kansas TR program include the right to kill one deer of a specific sex with a specific weapon. The permit gives a hunter the right to hunt, not the right to a deer. The right is valid only during the standard open season and only in the county of the landowner who originally obtained the right. Once sold, the right may not be transferred again. The deer permits are of limited duration; that is, they cannot be stored for use in future years. Nonresident hunters that purchase a TR in the Kansas program are required to comply with standard hunting regulations, hold a valid license to hunt in the state, and must hold documentation of their purchase of the permit during their hunt.

The allocation mechanism used in the Kansas program is worth noting. The permits are initially distributed through a random-draw lottery to resident landowners and land tenants/managers, each of whom may be granted up to one permit per year through the lottery. Winners in the lottery must pay the state a fixed fee for each permit they are allocated through the lottery. They then can resell the permits to nonresident hunters at market prices.

Based on a 2002 Kansas Department of Wildlife and Parks' survey of hunters, Taylor and Marsh (2003) report that the mean price of nonresident deer hunting permits for all hunting modes was US$760.13, which is much higher than the state's fixed price of US$205.50 for a permit in the lottery to landowners. However, the price nonresident hunters paid for permits was often bundled with other services including access to the land, guide services, etc. Using Taylor and Marsh's regression analysis evaluated at the mean of all variables other than guide services, we found that, on average, hunters who did not pay for guide services paid $656 for the permit. The permit price was reduced an additional $82 if they made use of land that did not require negotiating access with the landowner. The result was an average value of about $574 for the hunting rights alone. Hence, landowners who received permits through the lottery were able to earn substantial rents from the sale of permits to hunters.

There is some evidence that the situation in Kansas has changed. In 2006, the supply of permits for a number of regions in the state exceeded demand and surplus permits were made available after the lottery. These permits could be purchased from the government by residents or tenants at the fixed price of $322. At least for those regions, it is unlikely that permits would sell for much above this price so that the state appears to have captured a significant share of the rents that could be generated through these permits.

The transfer system for the Kansas hunting-rights program is also worth noting. The program is monitored by the state in that all permit transfers from landowners to hunters must be reported to the Kansas Department of Wildlife and Parks. This reporting can be done via the internet, so transaction costs do not appear to be substantial. There is also evidence that transaction costs associated with bargaining are dropping. Although only 2 percent of the licenses were purchased via the internet in the period studied by Taylor and Marsh, it is now possible to find many permits for sale on eBay, indicating that the market is becoming much more fluid.

The hunting programs reviewed here help us consider the possibilities for transferable rights in recreational fishing. And like in fishing, observing the level of participation in terms of days hunted, hunter success, where hunting takes place, etcetera of every participant is difficult and/or time consuming. On the other hand, recreational fisheries managers might also note that some hunting systems yield a great deal of information about hunters' participation. Hunter licenses provide a means of determining the number of hunters and deer tags indicate places and dates of kill (Leal and Grewell 1999).

The potential for raising revenue to fund fisheries management should not be overlooked. Policy makers are usually reluctant to create opportunities for profits from public fish and wildlife, but the Kansas program does just that. In the Kansas program, the government-set price for permits to landowners who draw them in the lottery has increased so that an increasing share of the rents is now captured by the state.

All of these programs are perhaps more politically palatable because they focus only on transferable nonresident hunting rights and they allow landowners in the state to benefit. No similar system exists for the resident permits. As such, one important drawback to this approach is that the effectiveness of these programs in controlling overall use of a resource is limited.

Critical Questions in Designing a TR Program

Based on our review of existing TR programs, we consider nine practical questions that must be answered as a necessary step in designing a TR program for a recreational fishery. Answers to these questions are not independent of one another, and two or more alternatives exist for each question. None of the questions has an obvious answer.

How Should Transferable Rights Be Measured?

The first question that must be answered is the unit of measurement of the transferable right. Recall that Dales (1968, 797) defined an asset unit as "the smallest physical amount of the asset to which it is practicable to apply property rights." What is the asset that is being used by recreational anglers: access to the recreational opportunity or the fish in the water? What is "practicable" to ration? The problem is further complicated when release mortality is high, since an angler's catch can lead to fishing mortality that is not easily monitored.

We see three alternatives for measuring a TR in a recreational fishery:

Alternative 1: Set the TR unit in number of fish caught (or retained).
Alternative 2: Set the TR unit in pounds of caught fish.
Alternative 3: Set the TR unit in fishing days.

In evaluating these three alternatives, we focus on Dales' advice that the rights be quantified in a way that is "practicable." There are three issues that must be considered: control over the biological impact of the fishing activity, scope of monitoring and enforcement, and level of transaction costs.[6] As an illustration, consider alternative 1. A single TR would grant its holder a right to harvest some number of fish, each one presumably of legal size. This approach has the most in common with the TR programs discussed above for hunting and fishing. For example, the TR unit for the Alaska charter-based IFQ program for halibut is specified in number of fish, and the TR unit for the Kansas hunting program is specified in terms of a single deer of a specific sex. Sutinen, Johnston, and Shaw (2002) point out that the rights granted to the recreational fishery in the Alaskan halibut program are denominated in fish and are as consistent as possible with existing regulations, which include daily bag and possession limits on numbers of fish. Thus, fishery regulators and biologists can focus on the same criteria, number of fish landed, as the means of harvest control. Although a permit based on number of fish relates quite well to the biological impact that anglers impose on the fish stock, some uncertainty would remain because of variation in the size and age of the fish that might be caught.

The biggest obstacles to adopting a fish-based TR in marine recreational fisheries are the problems related to monitoring and enforcement. As with bag limits, a right denominated in fish creates an incentive for fishers to discard smaller fish (that may not survive) in the hope of landing larger fish, so that the right is essentially used several times. It is unlikely that this is a major problem in big game hunting where discarding is relatively rare and more easily detected on land by wardens. But fishing is different in that it is harder to detect violations at sea. Johnston et al. (this volume) explore the use of tags in several recreational fisheries and argue that tags could be an effective way to enforce a TR program denominated in terms of a caught fish. In such programs anglers are required to terminate the right immediately after a fish is landed, usually by physically attaching the tag to the fish. Still, we are somewhat skeptical about the degree of compliance with such regulations in a deep sea fishery, where anglers regularly harvest multiple fish and multiple species on a single day. This may be less of a problem for charter boats, where randomly assigned and anonymous monitors could be used to increase compliance.

A fish-based TR that takes the form of a tag would need to be purchased in advance of a trip and would probably need to take a physical form such that it can be attached to the fish. This creates transaction costs for fishery managers and anglers because additional time and expense is required for tag acquisition. Further, because of uncertainty about how many fish an angler will land, anglers will either have a surplus number of tags before each trip, reselling unused ones or using them in the future if enough time is left in the season, or have too few tags and essentially face a smaller daily bag limit than necessary. These problems are substantially diminished in the case of charter boats since a charter boat operator could maintain a stock of tags for his or her clients and could use surplus tags on future trips.

Under alternative 2, rights would be stated in terms of a number of pounds. This approach would specify the TR in the same units as the official TAC, facilitating transferability of rights between commercial and recreational sectors and clarifying the fishing mortality associated with each TR used. The incentive to discard undersize fish would be reduced since a TR in pounds of fish use up less of the angler's right than a TR in numbers of fish. However, there are two problems with a pound-based right that would probably make it impractical. First, monitoring and enforcement would require that rights be purchased before a trip, resulting in a similar nuisance for nonchartered anglers as a TR denominated in numbers of fish. Second, unlike the one-fish, one-right approach under alternative 1, there is no obvious way that an angler would terminate a right upon landing fish that did not exactly match the allowed poundage. Transaction costs, therefore, are likely to be higher than those denominated in fish.

Under alternative 3, rights would be stated in terms of a number of days for fishing. Since fishing mortality depends on the angler's success, this approach is the least satisfactory in terms of its relationship between rights used and the physical impact on the fishery. If the TRs are denominated in terms of days of access to the fishery, then bag limits would probably need to be used to control total catch per day. As with TRs for nonpoint pollution, this alternative essentially represents a practice-based allocation, with its ensuant limitations. In this case an inefficiency arises because some anglers may not desire a complete bag limit while others may wish to exceed that limit. Pareto-improving trades between these two types of anglers are possible.

In spite of their known inefficiencies, practice-based TRs are used when monitoring actual environmental impacts is costly and incentives exist for noncompliance. Hence, this asset unit may be more appropriate for recreational fisheries with similar characteristics. As pointed out by the Scottish fishers interviewed by Rossiter and Stead (2003), rights defined in terms of days of access could also have advantages in terms of monitoring and enforcement. The day-based right could be enforced in much the same way as requirements that anglers have a fishing license. Although anglers know that it is unlikely that their license will be checked during a day of fishing, they usually purchase the license before fishing because of the uncertainty as to whether they will be caught. The day-based right might also be preferred because it is denominated in terms of the variable over which angler choices are most directly made—whether to go fishing or not. For ease of monitoring and enforcement, a user of a day-based right would be required to terminate that right prior to leaving the dock. Hence, rights could be tracked electronically as entries in a database identifying those anglers with a right to fish on a given day. The entire market could be embodied in an electronic tracking system in which an angler could purchase, sell, and terminate a right through the internet or a toll-free telephone number.

How Should Temporal and Spatial Elements of TRs Be Handled?

In specifying the units of the TR, their spatial and temporal characteristics must also be defined. Does a recreational TR grant its holder rights to use the fishery at any time and in any place, or in a limited region for a limited time period? First, we consider the spatial dimension with specific attention to a fishery with a predominant recreational sector, the red snapper fishery in the Gulf of Mexico. Would TRs be valid in all Gulf waters, or would they be partitioned into categories with each category of TR valid only for a specific area—state or even a smaller region? Based on the simplest conception of economic efficiency, economists would typically argue for no spatial limitation so that rights could go to those areas where the permits are most valuable. In the case of Kansas' TR hunting program and hunting programs in other states, however, spatial limits of permits are frequently imposed. There are two reasons that spatial limits might be imposed in a recreational fishery. One reason is that there may be equity considerations. For example, the vast majority of the permits might be purchased for use by anglers in a single region. Politicians in the remaining regions may lament the adverse impacts on the recreational sector of economies in the remaining regions or states and push for a fixed geographical allocations of TRs that ignore changing local conditions. A second reason for spatial restrictions is biological. As pointed out by Sutinen and Johnston (this volume), if rights became highly concentrated this could lead to localized stock depletion. We believe that the spatial scope of the TR should be determined primarily on biological considerations. Political considerations might also be considered, but this should be done with full awareness that efficiency costs will likely result.

With regard to the temporal dimension, there are two issues that must be resolved. First, would unused TRs expire at the end of the year? There is strong evidence that such expirations in TR programs can be counterproductive as they encourage use at the end of the year when delay would actually be preferred by some holders of those rights (Hahn and Hester 1989). Further, if a fishery's stock is below the maximum sustainable yield, then delaying harvest of a fish can only increase stock over time. Hence, automatic expiration of permits would be counterproductive in terms of the biological health of the fishery and the economic welfare of anglers. Despite these efficiency arguments, there is ample precedent for a time-limited right. Most fishing and hunting license and tag programs follow this approach. Unused fishing and hunting permits expire at the end of the season. IFQ percentage shares are perpetual in most commercial fisheries; however, the yearly catch allocations based on these shares are largely time-limited. For example, the recently approved IFQ program for the red snapper commercial fishery in the Gulf does not allow for unused shares in one year to be "banked" for later use, thus "removing financial incentives to not use shares" (Gulf of Mexico Fishery Management Council 2006, 145).

The second temporal issue is to control how long a permit would be valid for use. This is obvious if rights are specified in terms of a day of fishing. But if rights are specified in terms of numbers of fish or pounds of fish, then it is less clear without further definition. We see advantages to specifying the right as an ex ante right; that is, a right to catch some amount of fish, not a guarantee of actual possession. Under this approach, it might be possible for the rights to be tracked without the use of a physical tag, facilitating the electronic tracking and exchange of the TRs. On the other hand, in keeping with most hunting systems, some recreational fishing systems, and commercial IFQs, quantity-based rights would be valid until the fish are actually captured or until the end of the season.

Should Size and Bag Limits Be Retained?

As noted in the beginning of this chapter, size and bag limits can be very inefficient devices for the control of total harvests. Ideally, a TR program would replace these restrictions; however, there are legitimate reasons for maintaining restrictions. If a TR is defined as a day of use, then a bag limit would probably need to be retained to avoid giving individuals an incentive to harvest very high levels of fish on a single day. If rights are defined in terms of pounds or numbers of fish, then bag limits would seem unnecessary for controlling overall harvests. Nonetheless, in the proposed charter IFQs for Alaska halibut, in which rights are defined in terms of fish landed, a bag limit was retained. Similarly, a hunter who purchases a deer permit in the Kansas TR program faces the same seasonal bag restriction as other hunters in the state—one deer of a specific sex.

In light of the problems with size and bag limits discussed above, we believe that these management tools should be used sparingly as a means to reduce total catch. Regardless of the units in which the TR is established, size limits work well to protect particular age classes or help in the development of trophy-sized fish. In these capacities, they may play an important role in an efficient TR program, but for the purpose of reducing landings, size and bag limits can introduce economic inefficiencies.

How Should Monitoring and Enforcement Be Carried Out?

One of the major challenges of all TR programs is monitoring and enforcement. A TR program has value only to the extent the rights and obligations are enforced (King 2005). No matter how theoretically ideal an asset unit specification, if it cannot be enforced it will not adequately protect the resource or create value in the TRs. The traditional management tools of closures and gear restrictions are relatively easy to enforce. Fishing out of season or fishing with illegal equipment can be observed from spotter planes or from government vessels patrolling at sea. In contrast, particularly in a multiple-species, large scale, deep-sea fishery such as red snapper in the Gulf, it would be more challenging to ensure that all anglers fishing for snapper have the necessary right.

Comparing the different alternatives for measuring the asset discussed above, a day-based right may be easier to enforce since it could be monitored at the dock and would not require observation of all activities, including discards, while at sea. A program with fish- or pound-based permits, in contrast, could be circumvented by discarding fish for which the angler does not hold a right or by hiding one's catch.

A variety of incentives are created by a TR program. As the market price of the TR increases, so does the incentive to fish without the right.

Countervailing this, the fines for noncompliance might be tied to the price so as to ensure adequate disincentive to cheating. In addition, as noted by Scott (1988), a TR system creates a sense of common property among resource users, so that anglers who have purchased the necessary right might assist in watching for cheaters. Finally, if all or some of the TRs are sold by the government, the resulting revenue could be used to help cover the government's cost of monitoring. The funds available for monitoring would increase as the value of the TRs increases, setting in motion a process for strengthening monitoring and compliance and ensuring the integrity of the permits that were purchased.[7]

Sutinen and Johnston (this volume) point out that the wider the rights are distributed across regions and fishers, the more difficult monitoring and enforcement become. In any given recreational fishery, managers and participants are in the best position to determine which type of TR unit can be adequately policed. Keeping in mind that a TR program's effectiveness rests critically on the enforceability of the rights, consultation with users is a necessary step in the development of a recreational TR program.

How Should Transferable Rights Be Allocated Initially?

As exemplified by the failure to go forward with Alaskan halibut charter IFQs, the question of initial allocation is one of the most problematic issues to address in developing any TR program. Based on other TR programs involving public resources, the following are four alternatives for initial allocation:

Alternative 1: Grandfathering based on historical use
Alternative 2: Lottery
Alternative 3: Auction
Alternative 4: Federal sale (retail at fixed price)

Grandfathering is the most common allocation approach used in TR programs. In this case, TRs are initially allocated at no cost based on historical catch records of all eligible applicants. Such an approach has been commonly used in initial allocation of IFQs in commercial fisheries. Because grandfathering requires a foundation of credible records, this approach would be difficult to implement in recreational fisheries with a substantial number of anglers who fish on their own and could even be difficult in charter fisheries if operators have not been meticulous in their bookkeeping. The difficulty in establishing grandfathered rights is one of the reasons that Sutinen and Johnston (this volume) argue that rights should be issued to regional angling management organizations as discussed below.

Like grandfathering, if the initial allocation of TRs is established through a lottery, the rights would typically be given away at no cost. This approach is followed in many recreational systems in which the supply of available use rights is less than the demand. Applicants could apply separately, perhaps at a fee, for access rights, or all licensed anglers could automatically be qualified. In cases like the Gulf's red snapper fishery, where there is a substantial presence of charter and "party" boats, a separate allocation and lottery to those vessels might be needed to avoid a dispersion of the rights that allocates most to more numerous nonchartered anglers. Lotteries exist in a number of state big game programs—for example, Maine's moose hunting permit program, Kansas' nonresident deer hunting permit program, and New Mexico's lottery system for public elk hunters on private lands. Scrogin and Berrens (2003, 37) emphasize that "since lotteries ration independently of income, they are commonly favored by the public due to equity concerns." As in the Kansas nonresident hunting permit program, a lottery does not have to give the right to winners at no cost; winners may be required to pay a non-trivial price for the right.

Under alternative 3, the TR permits are distributed initially to potential users through an auction. In recreational fisheries, auction participants could include not only individual anglers, but also charter boat operators, groups of anglers, or retail vendors. Auctions are frequently used to transfer the use of potentially very valuable assets from public to private hands, such as timber rights on national forests and off-shore oil leases, and when the seller is unsure about the values that bidders are willing to pay. They have the advantage of making the process of allocation transparent—the winning bids are reported in news outlets, which is important in public-to-private transactions for resource use. Auctions can also be used to facilitate the purchase of multiple TRs by the for-hire sector since there need not be any limit on the number of rights a single buyer can purchase.

In the purely economic sense, using auctions to allocate initial TRs in fisheries is superior to other alternatives because it identifies potential fishers with the highest use value of the fisheries and maximizes revenues in an economically efficient way. Economic efficiency of the auction in the initial allocation of fisheries TRs, however, depends on the particular mechanism used in the auction. For example, if there are restrictions on who is eligible to bid—say, to protect a group of low-income fishers—then this mechanism makes the auction less efficient than one absent of such eligibility restrictions.

Morgan (1995) argued that the method of initially allocating commercial fisheries quotas will eventually move to auctions because quota allocation by administrative decision is economically inefficient. Notably, initial allocations of IFQs in yet-to-be-discovered commercial fisheries in New Zealand are now slated for auctions. This mechanism is not without controversy, however, as there would be protest that only the wealthiest fishers benefit and anglers may object to paying for access to a resource that they have historically used for free.

The final allocation option is to sell the licenses at a fixed price. For example, in the Gulf red snapper fishery, rights might be held throughout the year by the Gulf of Mexico Fishery Management Council or the National Marine Fisheries Service (NMFS) and sold directly to the public. A disadvantage is that, if the price chosen by the agency is too low, the fixed price could create opportunities for rent-seeking behavior on the part of fishing groups to capture the rents generated under TRs. In this case a secondary market would arise, creating opportunities for profiteering by those able to game the system and purchase their permits early. If a fixed-price approach is taken, governments may feel compelled to regulate transfers in a manner that diminishes the potential for the market to efficiently allocate the permits. A creative adaptation to this that could overcome many of the problems would be to have a government price that varies in response to supply and demand.

Who Should Receive Rights?

Related to the question of how rights should be allocated, we ask who should receive the rights. At first glance, one might assume that the answer to this question is obvious—rights should be held by anglers. On closer inspection, however, this is only one of several options. The Alaskan recreational halibut program, for example, was designed to distribute IFQs only to charter operators, leaving an unlimited number of non-chartered anglers to catch halibut. Any reduced harvest achieved by better control through TRs in the commercial and chartered recreational sectors can be compromised by the lack of effective control in the non-charter angling sector. If the size of this sector is significant, then managers may not be able to achieve sustainable overall harvests. A more comprehensive TR structure is proposed by Sutinen and Johnston (this volume) through so-called angling management organizations (AMOs). These organizations would be allocated TRs for management and distribution to anglers and possibly other recreational groups under their jurisdiction.

Listed below are these three alternatives and a fourth for consideration:

Alternative 1: Individual anglers
Alternative 2: Angling management organizations
Alternative 3: For-hire recreational sector only
Alternative 4: Local or regional governmental authorities

In evaluating the four alternatives, two important issues that must be addressed are the information that is available and the transaction costs that might result. As noted by Sutinen and Johnston (this volume), alternative 1 offers many advantages because of its close connection to the existing regulatory structure. However, they raise two main concerns: initial allocation issues and the question of enforcement. If rights are to be distributed initially using a grandfathering approach, this must be based on records of prior effort. However, most anglers obviously lack reliable documentation of such records. As noted above, monitoring and enforcement are challenges related to the definition of the asset unit. Nonetheless, if the TRs are distributed and held by private anglers, the problems of monitoring become much more complicated than if the rights were held by the for-hire sector or some other more manageably sized recreational structure than anglers themselves.

A structure favored by Sutinen and Johnston (this volume) is a collection of AMOs. According to their proposal, AMOs would be owned by anglers who would act like shareholders of a company. The organization, rather than individual anglers, would be granted the TRs, and the AMO could distribute rights to anglers. This could be done directly, or they could engage in more creative management options, such as arranging lotteries of rights, tournaments, etc. As the holder of the rights, the AMO would be responsible for monitoring harvests by anglers, and it would be the AMO rather than the anglers that would report actual harvests to the government. The authors believe that this could have economies of scale, reducing enforcement costs. Further, they argue that an AMO is also more likely to engage in activities that lead to improvements in the fishery, much like holders of IFQs sometimes participate in stock improvement activities (e.g., Repetto 2001). Sutinen and Johnston argue that the monitoring cost borne by government might be lower since the AMOs could be audited as a whole, rather than monitoring each angler's activities. The AMO option has not been fully critiqued, but Sharp (1998) notes that although AMOs were proposed for New Zealand, they were not adopted, in part because of its inconsistencies with legal authorities of management councils that currently oversee recreational fishing.

Alternative 3, in which the TRs would be held only by the for-hire recreational sector, is the system that was proposed for the Alaskan halibut fishery. This approach has advantages in terms of monitoring and because information for grandfathered allocations would probably be available. However, this would leave the private anglers out of the program, so it is only appropriate as a sole approach to a recreational fishery if the for-hire sector dominates the recreational fishery in terms of catch, now and in the foreseeable future.

Finally, under alternative 4 the TRs would be allocated to local or regional governmental authorities. These could function in much the same way as

AMOs, with some of the same advantages. However, the governments would, in turn, need to allocate the rights to anglers. Hence, this is not that much different from the status quo and not likely to yield improvements unless the regional authorities adopt some form of TR program themselves. This approach would also be undesirable for migratory species that cross regional jurisdictions; lack of enforcement by one governmental authority would have adverse consequences for the other regions.

Should Trades of TRs Be Monitored?

With the exception of products like military weapons and dangerous chemicals, most goods are transacted in markets without government monitoring of each trade. This is not true in many TR markets. TRs are essentially a government-granted right to exploit a public resource so it is usually the case that the government must be informed of TR trades. If monitoring were required, it would mean that all trades would need to be reported to a government agency. There is precedent for such monitoring in the various commercial IFQ programs at the federal fishery level and in the various state fish and wildlife programs. In case of the Kansas nonresident deer hunting permit program, all transfers are processed through the Kansas Department of Wildlife and Parks' main office. When trades are reported, it facilitates tracking of the rights and policing of the system. As is the case with many commercial IFQ reporting programs, where trades are reported electronically, the transaction costs can be kept low. Government tracking of all trades is required unless the right takes a physical form such as a card or a tag required in the angler's possession during fishing and not reusable physically once it is assigned to a landed fish. Notably, a deer tag in many western states requires the month and day of kill on the tag be punched out by the hunter immediately after the kill. In this case, care would need to be taken to avoid counterfeit tags, which could be costly.

Will Speculation Be Allowed?

Another question is whether individuals or firms should be allowed to buy and sell TRs to make a profit. Speculation in rights to fish and hunt is not typical. Fishing licenses are typically sold at a fixed price determined by the government with the seller usually receiving regulated issuance fees. The Kansas program allows the TRs to be sold at a profit, but they may be transferred only once. We consider here whether there is an economic justification for such restrictions.

In general, efficiency requires that the price be allowed to vary depending on supply and demand. Unrestricted trading can be problematic if there are concerns that individuals or firms might corner the market. In the particular case of the Gulf red snapper fishery, market power is likely to be a problem only if markets are isolated geographically—for example, if anglers have only one bait shop at which they can purchase their rights in a given town. Market design to reduce transaction costs so that markets do not become isolated would substantially diminish these concerns.

There are two other reasons restrictions might be sought. In a market occasional spikes in prices are possible, and this could lead to public outcry and opposition to the TR system. There will probably be public opposition if there is the impression that individuals are making substantial profits from a public resource. However, there is precedent for resale and speculation in the Kansas TR program. Additionally, the perception of inequality of access should not be entirely ignored. Once again, there may be public opposition to a program if prices become so high that only the very wealthy are able to fish.

Will Transfer between Sectors Be Allowed?

The basic economic notions of efficiency suggest that unfettered trading between the recreational and commercial sectors should be allowed (see Criddle, this volume). However the transfer of TRs between commercial sector and recreational sector can be controversial. While unconstrained transfer of the rights between sectors would increase short-run net benefits to market participants, it can have some negative consequences.

First, regional depletion could occur if the purchasing sector is geographically concentrated. Second, there is the potential for market concentration, particularly if rights are grandfathered and assigned in perpetuity as is done in most IFQ programs. The third reason for such a restriction would be concerns about secondary impacts on related economic participants. If the recreational sector purchased all the commercial rights, this could affect not only the fishers, but the processing and marketing sectors as well. Similar impacts on the tourism industry would occur if the trades went in the other direction. While in a full-employment economy such concerns have little economic merit, in situations of localized unemployment and/or situations with species-specific capital investments, such secondary impacts should not be ignored. Fourth, political pressure to protect the rights of one resource user group over those of another group may arise. Such pressures are likely to come with particular force from secondary market participants who have nothing to gain when rights are sold.

There is certainly a precedent for such restrictions. The proposed Alaska halibut program would have allowed charter boat operators to purchase IFQ shares from the commercial fishery, but shares originally allocated to the charter sector could not be sold to the commercial sector (Sutinen, Johnston, and Shaw 2002). Similarly, the Kansas program allows transfers only of the nonresident permits; resident permits are not transferable.

For the most part, economists favor market solutions in which free trade between sectors is allowed, with good reason (Aranson and Pearse, this volume). However, in developing a proposal for a TR program, economists should not ignore the real concerns that others will have about unfettered trade and the fact that allowing trade can actually be welfare decreasing in an economy that is not operating at or near a first-best equilibrium (Laffont 1988).

A Proposal for Recreational TRs in Gulf Red Snapper

Proposed here is a system for TRs for the Gulf of Mexico's recreational red snapper fishery that we believe would lead to substantial improvements in its management. The proposal is in three parts. The first addresses questions of the asset unit; the second responds to questions of initial allocation; and, the third pertains to various design issues of the TR system applied to this fishery.

Questions Related to the Asset Unit

The first consideration is to define the TR in such a way that its use correlates as directly as possible with the fishing impact on the resource. However, we return to Dales' point that the asset unit must be "practicable." A practicable TR is one that is easily monitored and enforced, with a design that keeps transaction costs at a minimum. It is possible that a fish-based right could achieve these standards; if so, it would be preferable to a day-based system.[8] But we have concerns that if the TR were specified in terms of fish caught, noncompliance could become a problem, trading could become difficult and costly to track, or both. With thousands of anglers participating across the Gulf and fishing miles out at sea, the ability to catch red snapper without actually using a tag would be great. Hence, our proposal is as follows:

1. TRs would be day-based rights. The rights would not take a physical form but would instead be a record on an electronically maintained registry. An angler would be required to terminate the right via the internet or a toll-free telephone call prior to beginning a fishing day. Such a program would be easily monitored and understood by anglers, making the market for TRs more credible, while keeping transaction costs very low.

2. TRs would not expire and would be valid to fish anywhere in the Gulf of Mexico. This would allow anglers the flexibility to avoid having to fish near the end of a season when they would prefer to wait until next season. In the event that excessive numbers of rights are carried over to future periods, the agency may choose to reduce the number of TRs issued in future years.
3. Since the right is denominated as a day fished, bag limits would be retained for the purpose of preventing excessive catch; but these limits could be increased, thereby giving anglers much less incentive to discard smaller fish. The combination of a day-based TR and bag limits also provides a means of controlling total harvests. Size limits should be used to address biological concerns regarding recruitment, but they should not be used as a device to reduce harvests.
4. Agency monitoring of all trades would be critical. This could easily be done through an electronic system that could be accessed via the internet or a toll-free telephone call. For example, a call to the database would allow an individual to transfer a TR to another angler by entering the fishing license or account number of the recipient. An internet interface could be designed that would not only transfer the right, but could carry out the monetary transfer as well.

This proposed system would track all fishing days and, when combined with dockside surveys, could yield very accurate estimates of the recreational catch throughout the season. This would be a great improvement over current data collection in the Gulf.

Initial Allocation Issues

As evidenced by the recent demise of the Alaskan charter halibut IFQ program, questions of initial allocation can be the most controversial and critical step in the development of a TR program. Although grandfathering is usually seen as the most politically palatable option in the commercial sector, it requires adequate records on an individual's catch history. Such records may be available from headboat and charter boat operators, but it is highly unlikely that private anglers would be keeping any records of historical catch that could be verified. It is possible that different allocation systems might be used, one for anglers and one for for-hire operators.[9] However this too might be controversial as it could create perceptions of unfairness. The option we advocate is to use auctions as a way to distribute the TRs. This may be controversial and politically unattractive, but it has advantages in terms of transparency and economic efficiency. Further, the resulting revenue could provide a source of funds for management, monitoring and enforcement, all of which would benefit the fishery in the long run.

Specifically, we propose:

1. The initial allocation would be carried out through multiple auctions during the year. A larger number of auctions might be necessary in the first years of the program, but after several years it may be possible to sell all rights in one or two auctions.
2. Rights could be bought by any individual, group, or for-hire operator.

If the complete auction of the TRs is politically infeasible, then separate allocations to some sectors could also work relatively well. For example, allocations to for-hire vessels could be made based on records of prior use and/or allocations could be distributed to regional groups based on estimates or prior landings.

Other Design Issues

There are other issues that need to be specified for TRs in this fishery. Given the electronically based market proposed, adequate monitoring and enforcement is critical to maintain system integrity. The monitoring agency needs to know which angler has the right to use which rights at any time. In addition, cheating must be effectively countered through the use of stiff fines for fishing without a valid TR.

No restrictions on speculation should be imposed so that the market will decide the clearing price of TRs. For example, a bait shop might purchase a large number of credits and then sell these off at a premium price or hold them in the expectation that the price will rise later in the year. Allowing this type of activity will create an economic incentive for the efficient spatial and temporal allocation of TRs so that the rights are moved to the time and place where they are valued most.

Initially, limited transfer between commercial and recreational sectors would be allowed with commercial rights in pounds converted to recreational rights using a conversion factor based on the bag limit and average weight per fish. A limit on the number of cross-sector trades might be imposed for the first five years of the program to mitigate concerns about secondary impacts and concentration. The motivation for allowing cross-sector trades is provided by Criddle (this volume) and Aranson and Pearse (this volume). In addition, this approach moves allocation decisions out of the political sector, which can be slow to react to changing fishery conditions, and into the marketplace where timely exchanges can be made as conditions change.

Conclusion

Despite the rapid expansion in the use of transferable rights as a tool for environmental and resource allocation problems, there has been relatively little use of this tool in recreational fisheries. Although there are many practical problems that must be resolved, we believe that they can be overcome. Similar problems have been faced in resource allocation problems ranging from water pollution to hunting. Creative solutions will be needed, but the current situation of declining seasons, increasing size limits, and decreasing bag limits is not tenable. A TR program offers hope as an alternative to this situation.

Bringing TRs into the recreational sector would break with tradition for managing recreational fisheries as states typically do not allow anglers or hunters to trade permits with one another, as opposed to landowners selling permits to hunters. There would also be challenges compared to applications of TRs in commercial fisheries, particularly in the scale of the effort since the number of participants in current transferable commercial IFQ systems range from one hundred or fewer fishers to a few thousand fishers (e.g., Alaska halibut). U.S. marine recreational fisheries like red snapper in the Gulf of Mexico can have tens of thousands of anglers in one region alone. Nonetheless, a TR approach has numerous advantages that make it worthy of serious consideration.

One critical question that we have not addressed here is the problem of bycatch and discard mortality. Anglers not targeting red snapper frequently catch snapper, and these anglers are unlikely to purchase TRs in a system directed only at red snapper anglers. The result will either be illegal harvests or discards, which will likely result in additional mortality. If targeting of the species is difficult and the problem of snapper bycatch and discard mortality in other fisheries is significant, the proposed TR system will not be adequate and alternative or additional management tools will need to be found.

With this caveat, we return to our initial question: Can a TR program work in a recreational fishery? We believe it can. There are numerous practical decisions that need to be made, and we offer possible answers to some of these above. Managers and stakeholders are in a better position than we are to determine what is practicable for any given fishery. Hence, our answers are proposed not as the final solution, but as a starting point from which managers, analysts, and stakeholders can start and look for alternatives that might be preferred.

Notes

1. The data presented include all Gulf of Mexico fishing trips. The 1997 MRFSS data include forty-four observations from anglers who targeted red snapper. Trips by these anglers are also distributed quite evenly throughout the year.

2. Estimates of release mortality are difficult to obtain, but there is some evidence that release mortality rates can be quite high, especially in deep water marine fisheries. See Harley, Millar, and McArdle (2000) and Burns, Koenig, and Coleman (2002).

3. An individual fishing quota (IFQ) program can also be called an individual quota (IQ) or individual transferable quota (ITQ) program.

4. See also NPFMC (2001a).

5. We draw here on information from web pages of the Kansas Department of Wildlife and Parks (http://www.kdwp.state.ks.us/news/hunting/big_game/deer.)

6. In Kim (2007) we carry out a theoretical comparison of the welfare consequences of alternative right specifications. In that analysis we find that there is no clear winner across the three alternatives.

7. Note that some level of monitoring is necessary to avoid an equilibrium in which the price is low because everyone is cheating.

8. We thank David Carter for suggesting the combination of daily bag limits and day-based rights.

9. Resources for the Future (RFF) economist James N. Sanchirico emphasized that each angler group might require separate consideration.

References

Arnason, Ragnar, and Peter H. Pearse. 2008. Allocation of Fishing Rights between Commercial and Recreational Fishers. This volume.

Batstone, Chris J., and Basil M. H. Sharp. 1999. New Zealand's Quota Management System: The First Ten Years. *Marine Policy* 23(2): 177–90.

Burns, Karen M., Chris Koenig, and Felicia C. Coleman. 2002. *Evaluation of Multiple Factors Involved in Release Mortality of Undersized Red Grouper, Gag, Red Snapper and Vermilion Snapper.* MARFIN Grant No. NA87FF0421. Paper presented at the Thirteenth Annual MARFIN Conference, Tampa, FL, January 16–17.

Clark, Colin W., Gordon R. Munro, and Ussif Rashid Sumaila. 2007. Buyback Subsidies, the Time Consistency Problem, and the ITQ Alternative Fisheries. *Land Economics* 83(1): 50–58.

Coleman, Felicia C., Will F. Figueira, Jeffrey S. Ueland, and Larry B. Crowder. 2004. The Impact of United States Recreational Fisheries on Marine Fish Populations. *Science* 305(5692): 1958–60.

Criddle, Keith R. 2008. Examining the Interface between Commercial Fishing and Sportfishing: A Property Rights Perspective. This volume.

Crocker, Thomas D. 1966. The Structuring of Atmospheric Pollution Control Systems. In *The Economics of Air Pollution*, ed. Harold Wolozin. New York: W. W. Norton & Co., 61–86.

Dales, John H. 1968. *Pollution, Property and Prices.* Toronto: University of Toronto Press.

Gulf of Mexico Fishery Management Council (GMFMC). 2006. *Final Amendment 26 to the Gulf of Mexico Reef Fish Fishery Management Plan to Establish a Red Snapper*

Individual Fishing Quota Program. (Accessed 7/5/2007 at http://www.gulfcouncil.org/Beta/GMFMCWeb/downloads/Amend26031606FINAL.pdf.)

Hahn, Robert W., and Gordon L. Hester. 1989. Marketable Permits: Lessons for Theory and Practice. *Ecology Law Quarterly* 16: 361–406.

Hanley, Nick, Jason F. Shogren, and Ben White. 1997. *Environmental Economics in Theory and Practice.* New York: Oxford University Press.

Harley, Shelton J., Russell B. Millar, and Brian H. McArdle. 2000. Estimating Unaccounted Fishing Mortality Using Selectivity Data: An Application in the Hauraki Gulf Snapper (*Pagrus auratus*) Fishery in New Zealand. *Fisheries Research* 45(2): 167–78.

Hood, Peter, and Phil Steele. 2004. History of Red Snapper Management in Federal Waters of the Waters of the U.S. Gulf of Mexico—1984–2004. *SEDAR7-DW-40.* St. Petersburg, FL: U.S. Department of Commerce, National Oceanic and Atmospheric Administration, National Marine Fisheries Service, Southeast Regional Office.

Johnston, Robert J., Daniel S. Holland, Vishwanie Maharaj, and Tammy Warner Campson. 2008. Fish Harvest Tags: An Attenuated Rights-Based Management Approach for Recreational Fisheries in the U.S. Gulf of Mexico. This volume.

Kim, Hwa Nyeon. 2007. Transferable Rights in a Recreational Fishery: An Application to the Red Snapper Fishery in the Gulf of Mexico. Ph.D. dissertation, Texas A&M University, College Station.

King, Dennis M. 2005. Crunch Time for Water Quality Trading. *Choices* 20(1): 71–76.

Laffont, Jean-Jacques. 1988. *Fundamentals of Public Economics.* Translated by John P. Bonin and Hélène Bonin. Cambridge, MA: MIT Press.

Leal, Donald R. 2002. *Fencing the Fishery: A Primer on Ending the Race for Fish.* Bozeman, MT: PERC. (Available at http://www.perc.org/pdf/guide_fish.pdf.)

Leal, Donald R., and J. Bishop Grewell. 1999. *Hunting for Habitat: A Practical Guide to State-Landowner Partnerships.* Bozeman, MT: PERC. (Available at http://www.perc.org/pdf/hfh.pdf.)

Leal, Donald R., Michael De Alessi, and Pamela Baker. 2006. *Governing U.S. Fisheries with IFQs: A Guide for Federal Policy Makers.* Bozeman, MT: PERC. (Available at http://www.ifqsforfisheries.org/pdf//ifq_governing.pdf.)

Morgan, Gary. R. 1995. Optimal Fisheries Quota Allocation under a Transferable Quota (TQ) Management System. *Marine Policy* 19(5): 379–90.

National Marine Fisheries Service (NMFS), Fisheries Statistics Division. 1997. *1997 Marine Recreational Fishery Statistics Survey.* Silver Spring, MD: NMFS. (Available at http://www.st.nmfs.gov/st1/recreational/queries/index.html.)

National Research Council (NRC). Committee to Review Individual Fishing Quotas. 1999. *Sharing the Fish: Toward a National Policy on Individual Fishing Quotas.* Washington, DC: National Academies Press.

North Pacific Fishery Management Council (NPFMC). 2001a. Final Motion on Halibut Charter Fishery Management April 14, 2001 (Draft). North Pacific Fishery Management Council Halibut Issues. (Available at http://www.fakr.noaa.gov/npfmc/current_issues/halibut_issues/401IFQmotion.pdf.)

———. 2001b. IFQ Implementation and Cost Recovery Committee December 2, 2001 Minutes. North Pacific Fishery Management Council Halibut Issues. (Available at http://www.fakr.noaa.gov/npfmc/current_issues/halibut_issues/IFQMinutes1201.pdf.)

———. 2003. IFQ Implementation and Cost Recovery Committee October 5, 2003 Minutes. North Pacific Fishery Management Council Halibut Issues. (Available at http://www.fakr.noaa.gov/npfmc/current_issues/halibut_issues/IFQImp100503.pdf.)

Pérez Henríquez, Blas. 2004. Information Technology: The Unsung Hero of Market-Based Environmental Policies. *Resources* 152: 9–12.

Repetto, Robert. 2001. A Natural Experiment in Fisheries Management. *Marine Policy* 25(4): 251–64.

Ribaudo, Marc O., Richard D. Horan, and Mark E. Smith. 1999. *Economics of Water Quality Protection from Nonpoint Sources.* Agricultural Economic Report No. 782, Economic Research Service, U.S. Department of Agriculture, Washington, DC.

Rossiter, Tom, and Selina Stead. 2003. Days at Sea: From the Fishers' Mouths. *Marine Policy* 27(3): 281–88.

Scott, Anthony D. 1988. Conceptual Origins of Rights Based Fishing. In *Rights Based Fishing*, ed. Philip A. Neher, Ragnar Arnason, and Nina Mollet. Dordrecht, The Netherlands: Kluwer Academic Publishers, 11–38.

Scrogin, David O., and Robert P. Berrens. 2003. Rationed Access and Welfare: The Case of Public Resource Lotteries. *Land Economics* 79(2): 137–48.

Sharp, Basil M. H. 1998. Integrating Recreational Fisheries into Rights Based Management Systems. Paper presented at the First World Congress of Environmental and Resource Economists, Venice, Italy, June 25–27.

Shotton, Ross, ed. 2001. Case Studies on the Allocation of Transferable Quota Rights in Fisheries. *FAO Fisheries Technical Paper,* No. 411. Rome: Food and Agriculture Organization of the United Nations.

Sutinen, Jon G., and Robert J. Johnston. 2008. Angling Management Organizations: Integrating the Recreational Sector into Fishery Management. This volume.

Sutinen, Jon G., Robert J. Johnston, and Reena Shaw. 2002. A Review of Recreational Fisheries in the U.S.: Implication of Rights Based Management. Working paper. Department of Agricultural and Resource Economics, University of Connecticut, Avery Point.

Taylor, Justin, and Thomas L. Marsh. 2003. Valuing Characteristics of Transferable Deer Hunting Permits in Kansas. Paper presented at the Western Agricultural Economics Association Annual Meeting, Denver, CO, July 11–15.

Woodward, Richard T. 2003. Lessons about Effluent Trading from a Single Trade. *Review of Agricultural Economics* 25(1): 235–45.

Woodward, Richard T., and Wade L. Griffin. 2003. Size and Bag Limits in Recreational Fisheries: Theoretical and Empirical Analysis. *Marine Resource Economics* 18(3): 239–62.

Woodward, Richard T., Ronald A. Kaiser, and Aaron-Marie B. Wicks. 2002. The Structure and Practice of Water Quality Trading Markets. *Journal of the American Water Resources Association* 38(4): 967–79.

Part II

INTEGRATING MANAGEMENT OF COMMERCIAL AND RECREATIONAL FISHING

Chapter 4

Allocation of Fishing Rights between Commercial and Recreational Fishers

Ragnar Arnason and Peter H. Pearse

The issue of allocation of fishing rights among sectors has recently begun to attract attention within fisheries communities and among policy makers, in our view largely as a result of the success of individual transferable quotas (ITQs) in generating substantial economic rents among commercial fishers. At least ten major fishing nations have adopted ITQs as their main, or at least a major, instrument for managing their fisheries, and between 10 and 15 percent of the global ocean catch is now taken under these arrangements (Arnason 2006b). These tradable rights greatly reduce the common property problem that has traditionally dissipated resource rents in open-access fisheries and enable fishers to organize their operations efficiently. Indeed, it can be demonstrated that appropriately designed individual quotas are capable of maximizing economic rents (Arnason 1990). This theoretical finding is supported by the successful experience of a growing number of fisheries around the world managed under individual quotas (Shotton 2000).

The encouraging experience with individual quotas as a means of improving resource utilization and generating economic benefits has led to suggestions that market trading in shares of the catch might be extended to deal with the problem of efficient allocation of catches among competing fishing sectors as well (Pearse 2006; Arnason 2006a). Just as the traditional common property problem in commercial fisheries arises from the negative externalities fishers impose on each other, competing recreational and commercial sectors impose externalities on one another. Similarly, the problem might be solved by creating property rights in harvests and enabling trade in these rights to allocate the catch among competing groups, resulting in substantial gain in social welfare.

This chapter deals with these issues. It is concerned mainly with the allocation between commercial and recreational fishers, although we briefly extend the discussion to aboriginal fishers and to passive or non-extractive users as well. Our focus is on the optimum allocation, which we define as the allocation that will maximize the aggregate net benefits generated by the resource to all users and the means of achieving that allocation.

We begin with a simple theoretical depiction of the optimum allocation of a harvest of fish between a commercial and a recreational group of fishers, both of whom seek to maximize their own benefits. We then examine the implications of the initial, regulatory allocation, leading to the conclusion that, no matter how much this differs from the optimum allocation, it will not prevent the optimum from being achieved as long as the rights are transferable without cost. However, if there are transaction costs, or other impediments to transferability, the optimum allocation cannot be expected to be achieved through trade in harvest rights. The potential efficiency loss is greatest when fishing rights are not tradable between sectors at all.

In addition, we discuss the practical difficulties of adjusting the allocation of fishing rights to the optimum through regulatory processes rather than relying on market trading. We also examine the special characteristics of other fishing sectors, notably aboriginal fishers and non-extractive users of fish. This leads us to a commentary on circumstances that call for sectoral organizations of fishers to represent them in allocation arrangements. The final section summarizes the conclusions of the paper and their implications for fisheries management.

Optimal Allocation of Resources: The Basic Principles

It is helpful to begin with a conceptual framework of the allocation problem and the optimum solution to it. For initial simplicity let us consider a stock of fish in a particular marine area, for which the fisheries authorities annually determine a total allowable catch shared by two groups—commercial and recreational fishers. The horizontal axis of the graph in figure 4.1 measures the range of possible catches up to the maximum available from the stock, or the total allowable catch, Q. The marginal (additional) benefit of an increment of catch over this possible range is measured on the left and right vertical axes for the commercial and recreational fishers, respectively, and is shown by the marginal benefit curve for each sector.

We assume, for present purposes, that the commercial sector consists of numerous small fishing enterprises facing perfectly competitive markets for their inputs and for the fish they produce in this fishery, so their aggregate benefit

Allocation of Fishing Rights between Commercial and Recreational Fishers 81

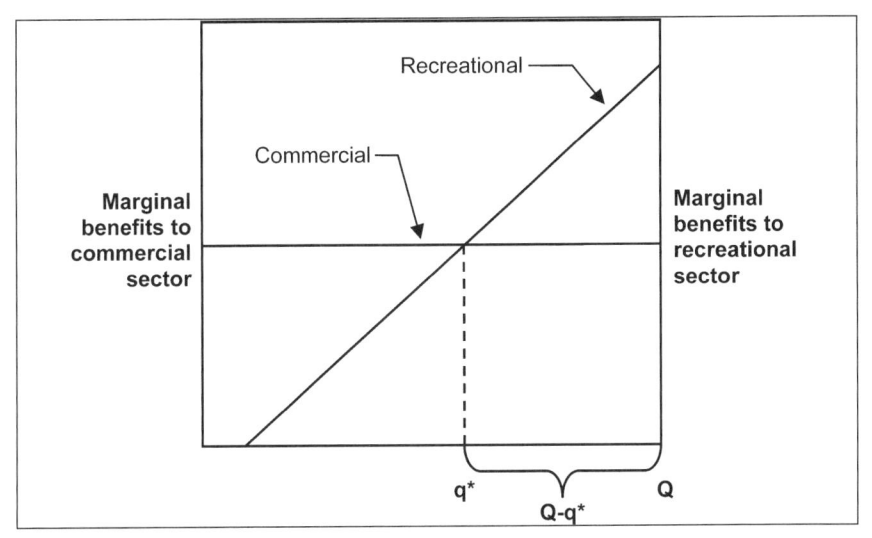

FIGURE 4.1
Optimal Allocation of the Catch (commercial and recreational sectors)

or profit function is drawn as a horizontal line. (Alternatively, we could allow for diminishing returns in the commercial sector and show it declining with increasing production.) As for the recreational fishers, we assume their marginal utility from fish caught diminishes with increasing catch, as reflected in the downward-sloping curve from the right-hand axis in figure 4.1.

For every possible level of catch allocated to either sector, its total benefit is reflected in the area under its marginal benefit curve up to that catch level. The objective is to allocate the catch between the two groups so that the aggregate social benefit of the resource, measured by the present value of the rights to all the catch, is maximized.

Obviously, in figure 4.1, the optimum allocation is a catch of q^* to the commercial sector and the remainder $Q-q^*$ to the recreational sector. This allocation equates the two sectors' marginal benefits and, consequently, maximizes total benefits, which are measured by the area under their combined marginal benefit curves. At any commercial allocation greater than q^*, the marginal recreational benefit exceeds the commercial benefit, so the total benefit can be increased by expanding the recreational share at the expense of the commercial, and vice versa for any commercial allocation less than q^*.

It should be noted that for many, perhaps most, commercially fished stocks, the marginal benefit to recreational fishers is lower than that of commercial fishers over the entire range of possible allocations. This type of situation is illustrated in figure 4.2.

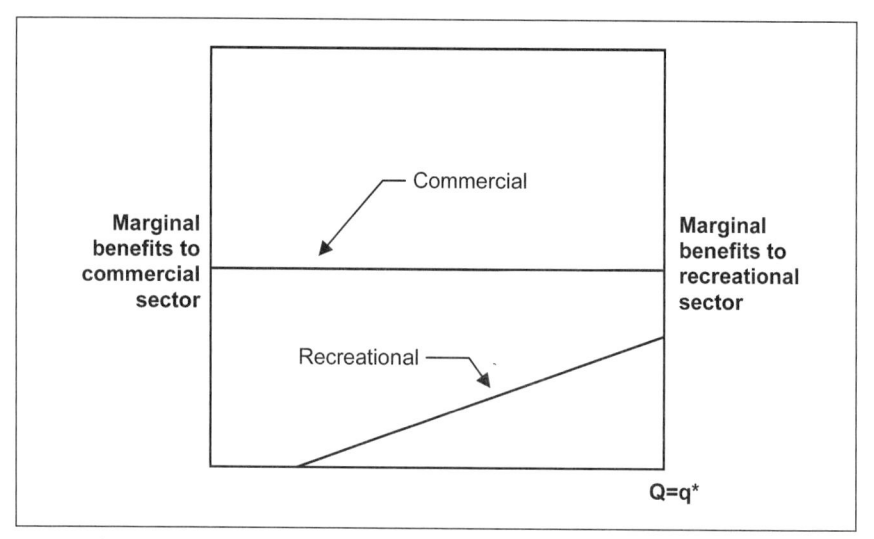

FIGURE 4.2
Optimal Allocation of the Catch (commercial sector only)

Obvious empirical examples are fisheries with no recreational fishing, suggesting the marginal benefits to recreational fishing are zero or negative. This seems, for instance, to be the case in many small pelagic (such as capelin and herring) fisheries around the world. In other fisheries, recreational fishing may coexist with commercial fishing while the highest marginal benefits to the former are less than to the latter. To establish empirical examples of this obviously requires substantial empirical research.

The reverse of the situation in figure 4.1, marginal recreational benefits exceeding marginal commercial benefits for all possible allocations, may be the case for certain stocks, such as Atlantic salmon, that are highly prized by recreational fishers relative to their commercial value. In those cases, social optimum requires allocating all the catch to the recreational sector. The optimum allocation involves sharing the catch only when each of the sectors has a marginal benefit exceeding the other over some range of potential allocations.

This simple analysis draws attention to certain fundamental aspects of the socially optimal allocation of fish and rights to fish. Most importantly, the optimal allocation depends on the marginal benefits of the parties involved and how these vary with the amount of their allocation. Two things follow immediately. First, the optimal allocation among fishing sectors cannot be determined from comparisons of the total benefits, or of the average benefits, gained by competing fishing groups. Thus, in debates about allocations between recreational and commercial fishers, the claim of either group that it generates the greatest value from the resource, even if true under the prevailing allocation, is of little relevance in determining the optimum allocation. Second, in order to work out the optimal allocation, the allocating authority has to know at least the relevant segments of the two marginal benefit curves. It is not sufficient to measure marginal benefits at certain points. This is a tall order for any centralized authority and probably not realistically feasible. It suggests the desirability of an automatic self-correcting mechanism that takes into account true marginal benefits to affect the allocations. One such mechanism is, of course, the market.

Regulatory vs. Market Methods of Allocation

Under suitable arrangements, a competitive market in rights to the available catch could be expected to allocate the fish optimally between the two groups. Consider the possibility that the commercial and recreational fishers in figure 4.1 are governed by a common individual quota system in which the rights to the catch are secure, exclusive, perpetual, perfectly divisible, and costlessly transferable. Assume that each group seeks to maximize the present value of its flow of benefits over time and that all prices, including the discount rate, are correct. Under these circumstances, if the allocation of fishing rights were at any point to the right of q^*, fishers in both groups could gain from transfers of fishing rights from commercial to recreational fishers. And if it were at any point to the left of q^*, they could benefit from transfers from recreational to commercial fishers. Thus trading would always tend to shift the division of the catch toward the optimal allocation. In practice, however, where harvests are shared by commercial and recreational fishers, the allocation between the sectors is typically determined by a governmental agency without much, if any, reference to the marginal value of catches to each group.

Moreover, any attempt to estimate the marginal benefit relationships would be fraught with severe difficulties of measurement. The products of commercial fishing are usually sold in competitive markets. Similarly, most of the inputs are bought in competitive and transparent markets. So, the marginal social valuation of most of the quantities involved is usually available. However, to estimate the aggregate marginal benefits of commercial fishing, it is necessary to obtain, in addition to these prices, estimates of the profit (or production or cost) functions of all current and potential fishers, which presents formidable difficulties.

For recreational fishing, measurement of the marginal benefit presents even more difficult empirical problems. Two difficult steps are involved. The first is to estimate the extent to which recreational fishing in a particular fishery would be affected by a marginal change in the quantity of fish allocated to this sector. A major complication arises from the product sought by recreational fishers; it is not simply (or even mainly) the fish harvested, but rather the recreational experience in which the fish caught are only one of many factors contributing to the quality and value of the experience, such as solitude, clean water, and good company (Canada Department of Fisheries and Oceans 1985). The variety of factors other than the catch that contribute to the recreation, their importance, and the way they vary in response to the catch allocated to this sector, present daunting analytical problems, closely related to those of joint consumption goods (Lancaster 1991). The second step involves estimating the value of this marginal increment of recreation. This task is complicated by the absence of market information; in North America, at least, recreational fishers usually have almost unlimited access to fishing opportunities, and to the fish, on public lands and waters without charge (apart from a nominal fee for a fishing license). Thus analysis of the value of recreational experiences must resort to indirect and inevitably somewhat tenuous techniques for estimating the consumer surplus fishers enjoy, or their hypothetical willingness to pay, based on inferences from their behavior, their expenditures on fishing, or their responses to hypothetical questions in surveys (Hanley, Shogren, and White 1997; Bateman and Willis 1995).

In addition to these challenges of measurement, the biological condition of fisheries, markets for fish, and economic and other circumstances of groups with demands on them continuously change, changing as well the optimum allocation, though governments rarely have the flexibility to respond to such fluctuations. For all these reasons, the usual recourse to governmental decision making is unlikely to achieve optimal allocations among competing groups of fishers, and allocations are likely to deviate substantially from the optimum.

As we have seen, market trading in fishing rights, in contrast, can be expected to lead to the optimum allocation without governmental intervention, and the process shortcuts the problem of estimating the marginal benefit functions of each sector (Arnason 1990).

The Initial Allocation: Does it Matter?

The market solution to this allocation problem, notwithstanding its apparent simplicity, also raises practical problems. Two requirements deserve attention here: the need to establish an allocation from which to start (which we refer to here as the "initial allocation"), and the need for transferability of fishing rights. The first of these is likely to be contentious. When individual quotas are introduced in commercial fisheries, the initial allocations are often the issue of sharpest disagreement (Shotton 2001), and it would no doubt be difficult in this context as well (though in many cases the already established catch shares of the competing sectors would probably narrow the scope for debate).

The need for a starting allocation also implies that the shortcomings and limitations of regulatory decision making, noted above, cannot be entirely avoided in adopting a market-based allocation policy. Because the government usually claims the rights of ownership of resources not held by private owners, a governmental process of some kind will be required to establish an initial allocation. This, as already discussed, is unlikely to correspond to the optimum allocation.

But does this matter? The answer is that it does not matter for efficiency as long as the allocated rights to the catch are transferable and there are no transaction costs. In these circumstances it will always be in the interests of the two groups of fishers to transfer rights until the optimal allocation is achieved.

If the regulatory authority imposed an allocation represented by a point to the left of q* in figure 4.1, where the marginal benefit of rights to the catch is higher for commercial than for recreational fishers, both parties could gain by transferring rights from the latter to the former, at some price between their marginal benefits. And if the initial allocation was to the right of q*, they could gain from transfers in the opposite direction. Clearly, these mutual trading gains vanish only at q*, the optimal allocation between the sectors. Thus market incentives, given free rein by costless transferability, would drive the allocation to the optimum. This establishes our first proposition:

Proposition 1: If fishing rights can be transferred freely—that is, without restriction and without cost—the allocation of actual catch will always shift to the optimal allocation.

Proposition 1 explains why economists tend to dismiss the problem of initial allocations as a distributional issue, of little lasting importance as long as the rights are transferable. But, as stated in proposition 1, this applies only under restrictive conditions, notably the absence of transaction costs or other impediments to trade.

Allocation with Transaction Costs

The absence of transaction costs is, of course, unrealistic. Normally buyers and sellers must incur costs to identify each other, to bargain and to arrange

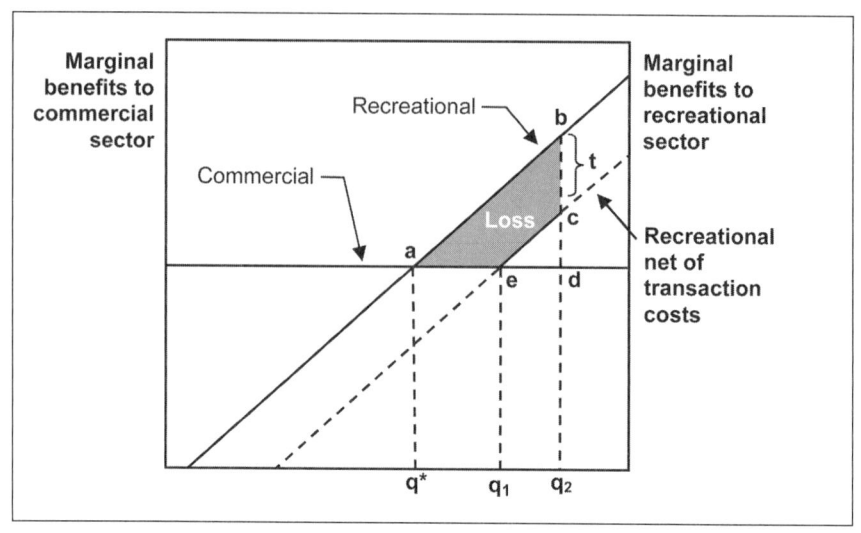

FIGURE 4.3
Market Equilibrium in Catch Allocation (with transaction costs)

transactions, and to pay any associated taxes and costs of enforcement and collection. So let us now consider the more realistic case in which transaction costs arise in transferring fishing rights. To simplify the exposition, assume the transaction costs consist of a fixed amount per unit traded (like a fixed commission fee, royalty or trading tax). Moreover, without loss in generality, let us assume that the buyer bears this cost, so his benefits from purchasing fishing rights are reduced by the amount of transaction costs. If he has to purchase all his resource rights, his marginal benefit curve is shifted uniformly downward, as illustrated in figure 4.3.

With reference to figure 4.3, suppose the commercial sector was allocated an amount of fishing rights q_2, exceeding its optimum allocation q^*. At that level, the marginal benefit of fishing rights is higher to the recreational sector, which would therefore be expected to purchase rights from the commercial sector. But because the recreationists' marginal benefits are reduced by the transaction costs (t in the diagram) the trading equilibrium is now at q_1, not q^*. In short, the existence of transaction costs prevents the market from reallocating fishing rights to the optimum. The resulting loss in economic efficiency, or benefits, is measured by the shaded area abce in figure 4.3. Part of this loss, the shaded area from q_1 to q_2, is eaten up by transaction costs. The remainder is the inability to get to the social optimum at q^*.

Thus, with transaction costs a heavier onus falls on the initial allocation because the market in resource rights cannot be relied upon to fully correct for deviations from the optimum. The resource will be used optimally only if the initial allocation is optimal, because only then will there be no need for market trading. Otherwise, trading would be needed to correct the misallocation, and this would trigger transaction costs, which would prevent the optimum from being achieved. This is our second proposition:

> Proposition 2: Transaction costs will prevent the market in resource rights from achieving the optimum allocation among user groups, leading to a continuing loss in economic benefits. The allocation of resource use will be optimal only if the initial allocation of rights is optimal.

Note that any initial allocation within the interval $[q^*, q_1]$, and a corresponding interval on the other side of q^*, will lead to no trading between the groups because the benefits to be gained fall short of the transaction costs. We refer to this interval around the optimum allocation as the "non-trading interval."

Figure 4.3 makes it apparent that the loss in efficiency depends on the amount of transaction costs, t, the shape of the two marginal benefit curves, and the initial allocation of rights relative to the optimum. If the allocation of rights is optimal, there will be no losses; if it is within the non-trading interval the loss will be less than that indicated in figure 4.3; and if the initial allocation is outside the non-trading interval the loss will equal that shown in figure 4.3.

In figure 4.4, we show a curve indicating how efficiency losses in the use of the resource depend on the initial allocation of rights for some level of transaction costs. As indicated by the curve, for an allocation close to the optimal use point, q^*, the efficiency losses are comparatively small. Then they increase rapidly and can become substantial. The non-trading interval depends, of course, on the transaction costs; the higher they are, the wider the non-trading interval. Also, with higher transaction costs the absolute slope of the curve increases, implying higher losses due to transaction costs for any allocation different from q^*.

Transferability

Fishing rights are often nontransferable, which can be viewed as equivalent to very high transaction costs. In terms of the above analysis, transaction costs high enough to eliminate trading would effectively extend the non-trading interval across the whole range of potential allocations. Accordingly, in the case of nontransferability, the allocation of fishing rights cannot deviate from the initial allocation.

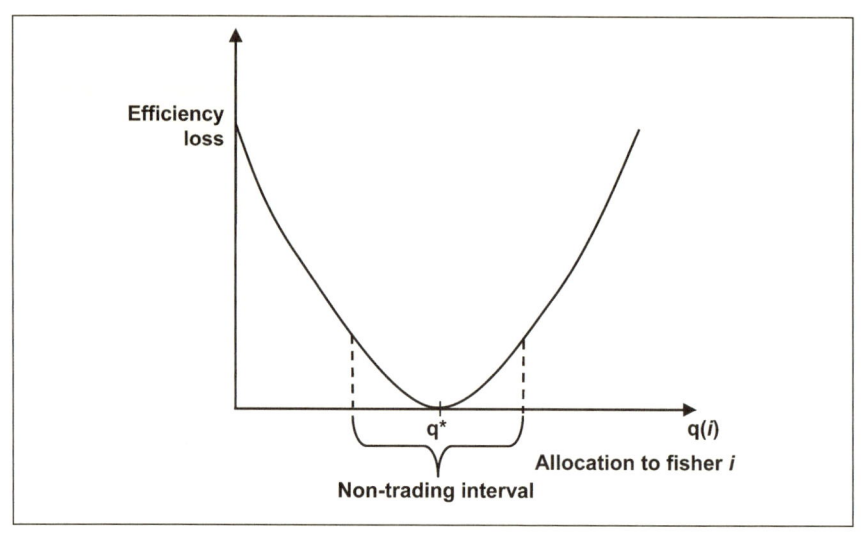

FIGURE 4.4
Transaction Costs: Allocation and Efficiency Loss

If the initial allocation is within the non-trading interval, the effect of nontransferability will be the same as that of transaction costs discussed in the previous section because no trading will occur in either case. But if the initial allocation is outside the non-trading interval, nontransferability will impose greater losses than transaction costs. In figure 4.3, the loss resulting from transaction costs is represented by the area abce, while that of nontransferability is abd.

In an extreme case, if rights are nontransferable and allocated to a sector that could not beneficially use them, all benefits would be lost permanently. Thus, if rights are nontransferable, the correct initial allocation is even more crucial than under moderate transaction costs.

Other Sectors and Organizational Issues

Fisheries often support more than two sectors. Allocation among three or more sectors cannot easily be depicted on a two-dimensional diagram as in figures 4.1 to 4.4. However, the conceptual nature of the problem does not depend on the number of sectors, and it can be analyzed by algebraic means. If a diagrammatic approach is preferred, the analysis can proceed in a succession of steps, each involving a diagram as in figure 4.3: first the allocation between one sector and all the rest taken together, then the allocation between a second sector and the then remaining group, and so on.

Two other sectors (or groups) deserve some comment in view of their importance in North American fisheries: (1) aboriginal fishers and (2) advocates of preserving fish stocks in a condition as close to their natural state as possible (to whom we will refer as "conservationists"). The curve of marginal benefit for both these groups can be expected to be downward sloping; the aboriginal fishers' marginal utility from an increment of fish for harvest will be less the more fish they already have allocated to them, and conservationists will put a lower value on protecting a fish stock from an increment of harvest the greater the proportion of the stock that is already protected from harvest.

Aboriginal people may fish under a variety of arrangements, including commercial and recreational fishing regimes, but we want to draw particular attention to the aboriginal right to fish for sustenance and traditional cultural purposes exercised by many First Nations in Canada and the United States. These rights are distinctive in two respects relevant to this discussion: they rest on a uniquely strong legal foundation (in Canada, being explicitly enshrined in the constitution), and they are not transferable (at least not to non-aboriginal people).

The nontransferability of aboriginal rights to fish means that neither market trading nor regulatory measures offer a means of adjusting the share of the catch provided to this sector in order to achieve an optimal allocation. The aboriginal entitlement is determined by other considerations, beyond the authority of fisheries managers and policy makers. Because of the strength of these rights, it can be considered for present purposes simply as an exogenously determined deduction from the total catch to be allocated among other sectors. The loss in efficiency, most of which is undoubtedly borne by the aboriginal people, is likely to be very substantial in some cases.

The conservation sector presents a unique problem insofar as the market for fishing rights does not offer a reliable means of allocating the catch optimally to this sector (Arnason 2006b). This is because two externalities are associated with purchases of fishing rights for conservation purposes. The first is the stock externality (Turvey 1964), which the ITQ system was designed to neutralize but is reintroduced by the holding of fishing rights for conservation purposes. A private trade of a fishing right from a fisher to one or more conservationists generates a positive externality, in the form of greater stock abundance and hence lower fishing costs to other fishers not party to the transaction. The opposite trade, from conservationists to fishers, generates a corresponding negative externality. The second externality arises from the public good character of conservation; conservation measures taken by one conservationist benefit all conservationists. Clearly, due to these two externalities, under decentralized trade in fishing quotas, individual conservationists will buy less fishing rights than would be socially optimal (Arnason 2006b).

Thus, in addition to the obstacle of transaction costs, there are two common circumstances in which individual suppliers and demanders in competitive markets for fishing rights cannot achieve the optimal allocation. One is where one or more sectors in the fishery do not employ tradable rights, such as individual transferable quotas, so individuals have nothing to buy and sell. The other is where conservation demand exists; in this case the actions of individual demanders will fall short of the optimum.

In both cases, the difficulty can be overcome by a representative organization of the individuals in the relevant interest groups, empowered to act on their behalf in holding their allocation, raising and spending money and buying quota from or selling it to fishers in other sectors. It can be demonstrated that such a representative body, trading in quota rights on behalf of the fishers in one sector with those in other sectors can, under an appropriate bargaining framework, lead to the optimal allocation (Arnason 2006b). This leads to our third proposition:

Proposition 3: If one or more sectors of a fishery are not organized under tradable fishing rights such as individual quotas, or if there is a demand for reduction of the harvest level below the regulated level, trade in fishing rights among individual suppliers and demanders cannot produce the optimal allocation. However, the optimum may be achieved through collective bargaining between the sectors.

Thus, with fishers and, for that matter, conservationists appropriately organized, rights to fish can be made transferable between sectors in a fishery to achieve overall efficiency even when the fisheries are not organized around individual quotas. The existence of individual quotas will, however, undoubtedly facilitate and encourage intersectoral transfers by enabling individual fishers to enjoy the benefits of trades.

One of the most conspicuous effects of adopting individual quotas as the means of allocating catches and organizing commercial fisheries, unforeseen in the early literature on the subject, has been the fading away of conflict over allocation not only among individual fishers but also among sectors of commercial fishers as well as between them and management authorities. Fishers, once they have a well-defined, secure, and long-lasting share of the catch, find themselves with a common interest in maximizing the overall value of the use of the resources. This not only creates an incentive for them to bargain with recreational fishers, conservationists and others for sharing the resources, but also an incentive to cooperate in data collection and research, stock enhancement, surveillance and enforcement, and other efforts to protect and increase the value of their resource rights (e.g., see Sharp 2005 and this volume;

Townsend 2005). A similar reliance on markets to deal with intersectional allocations can be expected to reduce conflicts between sectors and encourage cooperation among them.

However, recreational fishers and conservationists face formidable difficulties, both in adopting individual quotas and in creating representative organizations to act for them. In both of these sectors the participants are typically disparate, and they put a widely ranging value on fish. The identity and even the number of conservationists are unknown, as is the case with recreational fishers where they are not licensed. In contrast, aboriginal groups and commercial fishers are often already well organized to act collectively, and many have adopted individual quotas.

A Case in Point

To our knowledge, there are very few examples of market trading in fishing rights to adjust the allocation of catches between commercial and recreational sectors. Recreational fishing organizations in Iceland have bought up farmers' traditional netting rights in salmon rivers, and in the North Atlantic, the North Atlantic Salmon Fund (NASF), which represents the interests of recreational salmon fishers, has raised substantial funds to buy ocean salmon fishing rights in the North Atlantic from commercial fishers (McBride 2005). A particularly interesting example, which illustrates some of the possibilities and challenges we have discussed above, is developing in Canada's Pacific halibut fishery, where a commercial sector using individual quotas shares the catch with recreational fishers.

Canada's Pacific halibut fishery is a large, mature fishery dominated by a commercial sector organized under individual quotas. By the year 2000, the recreational catch had grown to about 9 percent of the total allowable catch. In response to growing anxiety among commercial quota holders about the encroachment of sportfishing on their share of the allowable catch, the Minister of Fisheries assigned the recreational sector a "cap" of 12 percent, giving it room to grow. He also declared that if recreational fishers wanted to increase their share beyond that level in the future he would expect them to acquire it through a market mechanism. Further, he announced that if their catch continued to grow and they failed to correspondingly expand their entitlement by purchasing quota, he would impose regulatory measures on recreational fishing to constrain their catches to their authorized limit.

Meanwhile, the recreational sector's initial allocation has exceeded its catch, and the commercial sector has "leased" the recreational sector's surplus in return for cash payments based on the price of quota in the commercial sector. By the end of 2005, payments for the recreational sector's surpluses had resulted in a fund approaching two million Canadian dollars. Because of the continuing growth in the recreational catch, the Department estimates that, in 2006, it will reach the recreational sector's 12 percent allocation. Any further expansion will require the recreational sector to buy some of the rights of commercial fishers.

A major difficulty in the development of this intersectional rationalization has been the absence of an organization of the recreational halibut fishers capable of representing them, bargaining, holding money and dealing in fishing rights on their behalf. The Department of Fisheries and Oceans having left the two interest groups to work out feasible arrangements, the organization of commercial fishers has had to deal with a sportfishing advisory board appointed by the Minister. This board lacks the authority to represent recreational halibut fishers or to hold money, a problem dealt with by the expedient of a trust fund established by the commercial organization in favor of recreational fishers.[1]

This innovative development provides an illustration of several issues discussed in this paper. First, it corroborates our characterization of sectoral benefit functions in figures 4.1 to 4.3. Recreational fishers clearly ascribe a positive value to their catch of halibut. However, their uncaught allocation during the early years shows that their marginal valuation is downward-sloping.

Second, market forces will tend to shift the allocation of the catch toward the optimum allocation by transferring rights to the catch from the sector in which its marginal benefit is lower to that in which it is higher. In this case, when the recreational sector's marginal benefit was zero and the commercial sector's was significantly positive (as reflected in the price of commercial quota), trading shifted resources to the commercial sector. It appears, also, that these intersectional transfers were made with modest transaction costs, simply utilizing the established and efficient mechanism for trading in commercial quotas.

Third, this case illustrates the relative ease with which an established commercial sector can organize itself to accommodate market trading in fishing rights, including mutually beneficial intersectional trading, in comparison with the recreational sector. Indeed, it suggests that the difficulty of organizing recreational fishers may prove to be the biggest obstacle to intersectional trade.

Finally, though not documented above, this experience has revealed the importance to the recreational sector of certainty of fishing opportunities and the comparatively high cost recreational fishers face in adjusting to fluctuations in their available catch on relatively short notice. Thus, while commercial fishers seem content to increase or reduce their catches providing their adjustments are compensated through market transactions, recreational fishers cannot easily accommodate such changes. Especially for commercial-recreational demanders, such as charter boat enterprises, fishing lodges, and guides, but also for others simply planning fishing vacations, any prospect of the fishery having to be closed or restricted because the sector has exhausted its allocation threatens a major disruption. This difficulty is exacerbated, of course, wherever sportfishers cannot acquire individual quotas to protect themselves from such uncertainty.

The importance to recreational fishers of stability and certainty of their fishing opportunities is likely to lead them to buy more rights from commercial fishers when the allowable catch must be reduced and to sell them when it is increased, leaving commercial fishers to make the needed adjustments in catches in exchange for cash payments.

Conclusion

In this chapter we have examined the conditions for optimality in the distribution of the catch among sectors in a fishery—that is, the distribution that will maximize the aggregate benefit generated by the fish resources. We have also restated the well-known economic theorem that any misallocations will be corrected by perfect markets. But we draw attention to the empirically common transaction costs and sometimes man-made restrictions on the transfer of fishing rights that will prevent markets from achieving optimal allocations. These market imperfections put a heavier onus on the regulated starting positions of the sectors; the greater the transaction costs and restrictions on transferability, the greater the loss from imperfect allocation in the first place.

In many fisheries, one or more sectors do not have quantitatively defined fishing rights, so individual fishers have nothing to trade. In such cases they need representative organizations to act on their collective behalf in the trading of fishing rights with other sectors. When there is significant demand for conservation of fish stocks, individual conservation activities generate special externalities that also call for collective action through representative organizations.

The chapter leads to the conclusion that fisheries policy makers concerned to achieve the most beneficial allocation of catches among fishing sectors should take care to:

- Ensure that the fishing rights allocated to all sectors have well-developed characteristics of property and, in particular, are freely transferable.

- Minimize transaction costs, or at least avoid increasing them with unnecessary regulations.
- Make their initial regulated allocations as close to optimal level as possible.
- Encourage and facilitate the organization of the fishers in each sector and other bona fide stakeholders in the fish stocks and their environment.

Notes

1. Since this was written preliminary data suggests that growth in the recreational catch and reduction of the total Canadian TAC for halibut have combined to push the recreational catch above its 12 percent "cap" in 2006 and 2007 (though the catch estimates may yet be revised in light of new data). Significantly, representatives of the recreational sector have begun to investigate ways of reconciling their need for an increased share of the catch with the commercial individual quota system. Equally notable is a new effort on the part of the Department of Fisheries and Oceans to facilitate this policy development; it has established a working group, with representatives of both sectors and an independent facilitator, to organize the collection of data, measures to contain recreational catches within that sector's allocation, a market process for transferring catch shares between the sectors, and a body empowered to act on behalf of the recreational fishers in raising, holding, and expending funds.

References

Arnason, Ragnar. 1990. Minimum Information Management in Fisheries. *Canadian Journal of Economics* 23: 630–53.
———. 2006a. Commercial Allocation Issues. Paper presented at the Sharing the Fish 2006 conference, Perth, Western Australia, February 26–March 2.
———. 2006b. Conflicting Uses of Marine Resources: Can ITQs Promote an Efficient Solution? *Institute of Economic Studies Working Paper Series* W06:07, Reykjavík: University of Iceland, December. (Available at http://www.ioes.hi.is/rammi32.html.)
Bateman, Ian J., and Kenneth G. Willis, eds. 1995. *Valuing Environmental Preferences: Theory and Practice of the Contingent Valuation Method in the US, EU and Developing Countries*. New York: Oxford University Press.
Canada. Department of Fisheries and Oceans. Economic Analysis and Statistics Division, Surveys Unit. 1985. Sport Fishing in Canada 1985. Ottawa: Department of Fisheries and Oceans.
Hanley, Nick, Jason F. Shogren, and Ben White. 1997. *Environmental Economics in Theory and Practice*. Macmillan Text in Economics. Houndmills, Basingstoke, Hampshire, England: Macmillan Press Ltd.
Lancaster, Kelvin J. 1991. *Modern Consumer Theory*. Aldershot, Hants, England: Edward Elgar.

McBride, Stewart. 2005. Icelander in Fight to Save Salmon: Wading into the Bureaucracy of Fishing. *International Herald Tribune*, June 17. (Available at http://www.iht.com/articles/2005/06/16/features/salmon.php.)

Pearse, Peter H. 2006. Allocation of Catches among Fishing Sectors. Paper prepared for the Sharing the Fish 2006 conference, Perth, Western Australia, February 26–March 2.

Sharp, Basil M. H. 2005. ITQs and Beyond in New Zealand Fisheries. In *Evolving Property Rights in Marine Fisheries*, ed. Donald R. Leal. Lanham, MD: Rowman & Littlefield, 193–211.

———. 2008. Recreational Fishing and New Zealand's Evolving Rights-Based System of Management. This volume.

Shotton, Ross, ed. 2000. *Use of Property Rights in Fisheries Management.* Proceedings of the FishRights99 Conference. Fremantle, Western Australia, November 11–19, 1999. *FAO Fisheries Technical Paper* 404/1 and 404/2. Rome: Food and Agriculture Organization of the United Nations.

———, ed. 2001. Case Studies on the Allocation of Transferable Quota Rights in Fisheries. *FAO Fisheries Technical Paper* 411. Rome: Food and Agriculture Organization of the United Nations.

Townsend, Ralph E. 2005. Producer Organizations and Agreements in Fisheries: Integrating Regulation and Cosean Bargaining. In *Evolving Property Rights in Marine Fisheries*, ed. Donald R. Leal. Lanham, MD: Rowman & Littlefield, 127–48.

Turvey, Ralph. 1964. Optimization and Suboptimization in Fishery Regulations. *American Economic Review* 54: 64–76.

Chapter 5

Harmonizing Recreational and Commercial Fisheries: An Integrated Rights-Based Approach

Ragnar Arnason

When one way of using a resource reduces the benefits from using it another way, the two uses may be said to be conflicting. Cases of conflicting resource use are common. One is the use of fish stocks for commercial fishing on the one hand and recreational fishing on the other.

Since the 1776 publication of Adam Smith's work (1977 ed.), it has been known that, under a complete system of property rights and smoothly functioning markets, market trades will normally bring about a Pareto-efficient solution[1] to conflicting uses of resources. However, in a situation of limited property rights, not to mention the common property arrangement—both of which are typical in fisheries—the outcome will generally not be anywhere close to being Pareto efficient.

In recent years, various forms of property rights have been implemented in ocean fisheries in order to alleviate the common property problem in commercial fisheries (NRC 1999; Shotton 2000). The most widely used of these are individual harvesting rights usually referred to as individual quotas (IQs). When these quotas are transferable they are referred to as individual transferable quotas (ITQs). It has been shown that under fairly unrestrictive circumstances, appropriately designed ITQs are capable of maximizing economic rents from a fishery (e.g., Arnason 1990). This theoretical result has received empirical support from a number of

---

I would like to thank the editors of this volume for extremely helpful comments on an earlier version of this chapter.

ITQ-managed commercial fisheries around the world (OECD 1997; Shotton, 2000; Hatcher et al. 2002).

The benefit of using ITQs in commercial fisheries raises the question whether a similar arrangement is capable of resolving the conflict between commercial and recreational fishing in a Pareto-efficient manner. This chapter attempts to provide an answer to this question. The approach is analytical. A simple model of the fishery including both commercial and recreational fishing is specified. With the help of this model, I investigate whether an integrated ITQ system (i.e., one encompassing both recreational and commercial fishers) is capable of efficiently resolving the problem of conflicting uses of a resource. The answer is yes. By contrast, it is found that a nonintegrated ITQ-system (i.e., one where the two groups of users are on separate ITQ systems) is not capable of efficiently resolving the resource use conflict.

The rest of the chapter is organized as follows. I begin by describing a particular case of market-based coordination of commercial and recreational use in a fishery, the North Atlantic salmon fishery. From that empirical example, I proceed to outline the theoretical model components, which are very simple. On the basis of that model, I then show how the ITQ system can resolve the conflict within the group of commercial fishers. Then, I establish the optimal joint utilization of the resource and show that an integrated ITQ system (i.e., an ITQ system that incorporates both commercial and recreational fisheries) is capable of replicating that solution. I go on to explore the role of management system integration in this result and show that nonintegrated ITQ systems, or for that matter any other nonintegrated systems, will not in general solve the problem. I conclude with a summary and discussion of some practical aspects of an integrated commercial–recreational ITQ system.

The North Atlantic Salmon Fishery

Atlantic salmon (*salmon salar*) is one of the most valuable (per unit weight) species of fish in the North Atlantic. Like other species of wild salmon, Atlantic salmon is anadromous, spending the early part of its life cycle in fresh water and the maturing part, when it undergoes most of its growth, in saltwater. During its saltwater phase, Atlantic salmon ranges all over the North Atlantic from the Gulf of Maine and North Labrador in the west to the Kola Peninsula and Bay of Biscay in the east. Historically, before the current depressed state of the stocks, most of the rivers around the North Atlantic supported stocks of salmon.

Atlantic salmon becomes primarily susceptible to human harvesting at two stages during its life cycle. The first is when the salmon congregate on certain favored feeding grounds in the ocean. The second stage is during their spawning runs when they appear in mass at river mouths and in rivers.

Harvesting salmon from spawning runs in rivers is a very ancient form of economic production (Lackey 2005). Since salmon are comparatively easily caught during their spawning runs and, therefore, highly susceptible to overexploitation, social institutions to control the harvesting emerged very early in history.[2] Most often, these controls were based on local land ownership and, in Northern Europe at least, they have for the most part persisted to the present.[3] These arrangements, by and large, proved sufficient to deal with the rising demand for recreational salmon fishing in the twentieth century. River owners simply shifted from traditional harvesting of salmon to the more profitable activity of selling fishing licenses or harvesting rights to recreational fishers.[4] This relatively smooth transition from one use to another illustrates some of the principles discussed in this chapter.

By 1960, however, ocean fishing and fish-finding technology had developed to the point where ocean fishing for salmon had become profitable. The ocean fishery was not subject to private property rights or other constraining social institutions of significance. Thus, it constituted a common property fishery. The common property problem was exacerbated by the fact that fishing fleets from several nations, some with no salmon stocks of their own, were involved in the fishery. Needless to say the fishery expanded fast. Soon spawning runs to the rivers began to falter—in some rivers they all but disappeared. By the mid-1970s, even the ocean fishery began to contract. The evolution of salmon harvests is illustrated in figure 5.1. As shown, it peaked in 1974; since then, it has greatly declined. No doubt the decline in stocks more than matches the decline in harvests. The decline was not halted until the latter half of the 1990s. Since then, the harvests have remained fairly stable at just under 3,000 metric tons per year, a quarter of the harvest in the early 1970s and about half the annual harvest before the great expansion of the ocean fishery.

So, in the 1980s, the North Atlantic salmon fishery situation was broadly as follows: Firstly, there was an extremely valuable recreational salmon fishery in rivers using rods. Secondly, there were still substantial remains of a traditional commercial salmon fishery primarily in and around river mouths using nets and weirs. Thirdly, there was the recently developed international, commercial ocean fishery employing driftnets and longline. The last was by far the largest in terms of volume of harvest. In terms of value, however, it was just a fraction of the much smaller recreational fishery. By this time, moreover, all three fisheries, especially the river fisheries, were rapidly declining as illustrated in figure 5.1.

It was into this difficult situation, with national governments apparently unable to find a solution, that the North Atlantic Salmon Fund (NASF)

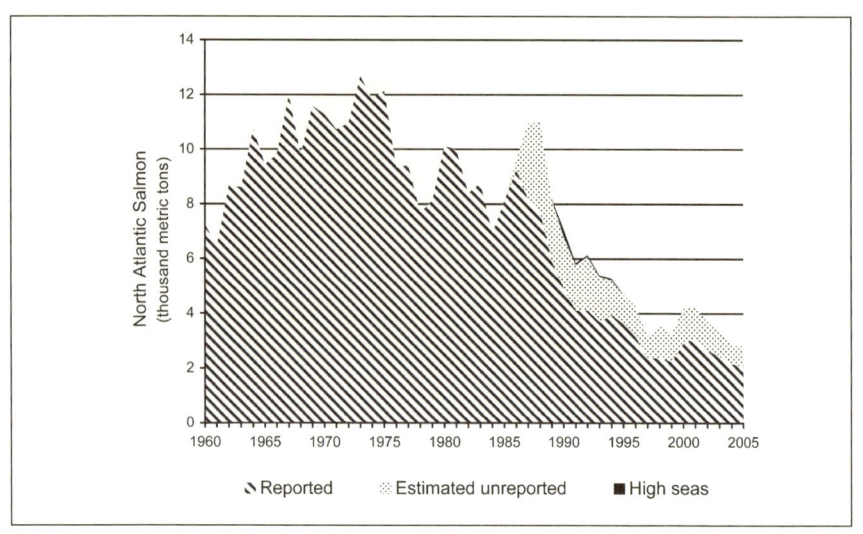

FIGURE 5.1
North Atlantic Salmon Catch, 1960–2005
Source: ICES (2007, 34).

entered. The NASF was established in 1989 by an Icelandic businessman, Mr. Orri Vigfusson, a passionate salmon sport fisherman. The co-founders of NASF were other recreational salmon fishers and stakeholders in salmon rivers concerned about the decline in natural salmon stocks. The NASF is set up as a private nonprofit organization. Its aim is to eliminate commercial fishing for North Atlantic salmon and, thus, protect and, hopefully, restore natural salmon stocks in rivers. In this respect, the NASF is similar to many other environmental organizations. Its chosen method for achieving this aim, however, is a bit unusual. Its method is not so much to convince governments to outlaw commercial fishing for salmon but to negotiate with and purchase fishing rights from commercial fishers. Thus, the NASF philosophy is to respect individual rights, even when these rights are poorly defined and diffuse, and to achieve its aims by voluntary market transactions. In this sense, the NASF provides a real life example of the principles discussed in this chapter.

The NASF has been remarkably successful. To date it has raised over US$35 million, about two-thirds of which are private donations and one-third public funds awarded by governments. These funds have been used to purchase, at least temporarily, most ocean and river mouth fishing rights around the North Atlantic. The remaining ocean fishing for salmon consists of about 40 driftnet rights off Ireland and northern England, some commercial, coastal-based gillnetting in the White Sea, Norwegian and

Scottish fjords, as well as a few native subsistence gill nets in the Gulf of St. Lawrence. Thus, thanks to the activities of the NASF, the ocean fishery for North Atlantic salmon has been greatly reduced. At the same time, whether or not due to the reduction in the ocean fishery, the decline in salmon catches has halted (see figure 5.1). This has greatly benefited the recreational fishery and, almost certainly, hugely increased the overall value of the salmon fishery.

This success is all the more impressive for the fact that the NASF has had to operate in an environment of ill-defined property rights and poorly organized or unorganized rights holders. It seems obvious that NASF's job would have been much easier with well-defined property rights such as individual quotas or similar. The lesson seems to be that even in a situation of a high number of poorly organized holders of quite weak forms of property rights, a scope for a negotiated solution exists and is attainable, provided the overall benefits of such a solution are great enough.

Basic Modeling Components

Consider the situation where a fish stock has two different uses: (i) the generation of commercial harvests and (ii) the generation of recreational fishing pleasure. To the extent that both activities affect stock growth and the benefits depend on the size of the fish stock, these uses conflict. Typically, harvesting by commercial fishers will reduce benefits to recreational fishers, and vice versa. This conflict in resource use is not restricted to the commercial and recreational fishers as groups. Harvesting by an individual commercial fisher will generally negatively affect the opportunities of all other commercial fishers and, by inference, all recreational fishers (Gordon 1954). Similarly, harvesting by an individual recreational fisher will negatively affect all other fishers, recreational and commercial.

Let the fish stock evolve according to the equation:

(1) $\dot{x} = G(x) - z$,

where x represents the stock biomass and z the aggregate harvest. The function $G(x)$ is the biomass growth function exhibiting the usual properties (see, e.g., Clark 1976).

Consider now a commercial fishing industry with an instantaneous benefit (profit) function:

(2) $\Pi(q,x)$,

where q represents the commercial harvest. The function $\Pi(q,x)$ is assumed to be monotonically increasing in biomass, x, and concave in both arguments. To make the situation interesting, we assume that there exists a biomass level such that the benefit function is positive and increasing in the harvest level, q, up to a certain point. The fishing industry consists, of course, of a number of different fishers.

Next consider a recreational fishery. For our purposes it is immaterial whether this is a commercial recreational fishery, where specialized firms offer fishing recreation to customers, or a pure recreational activity, where individuals simply fish for their own enjoyment, or a combination of both. Let the benefit function in this fishery be:

(3) $U(y,x)$,

where y represents the recreational harvest. This function, although different from that of the commercial fishery, is assumed to have the same basic shape. It is monotonically increasing in biomass x, increasing in the recreational harvest level y, at least up to a point, and concave in both arguments. Note that the function $U(.,.)$, may be regarded as the (aggregate) utility function of the recreational fishers.

As already stated, both groups, the commercial fishers and the recreational fishers, consist of a number of individual members. The preferred fisheries policies of these members will generally not be identical. Thus, within each group there will be internal conflicts just as between the groups. It does not pose any analytical problem to model this explicitly. In this paper, however, in order to focus on the fundamental conflict between commercial and recreational fishers as groups, we ignore this aspect of the situation and proceed as if both parties act as single units.

The above model is very simple. Although it certainly captures certain key elements of commercial and recreational fisheries, it ignores many of the real-life complexities. Nevertheless, I believe this approach is useful. The purpose of the exercise is to check whether certain property rights-based management systems can generate economic efficiency. A management system that does not work in the simplified context is very unlikely to work in a more complicated setting. By the same token, only systems that work in the simplified framework may work in a more general situation.

What Fisheries Policy Do the Parties Want?

We take it for granted that each party seeks to maximize the present value of their flow of benefits over time. For simplicity we assume, moreover, that all have the same rate of discount, r.[5] The two groups respective maximization problems are:

Commercial Fishers

Commercial fishers seek to solve the following maximization problem:

(I) $\quad \underset{q,y}{Max} \int_0^\infty \Pi(q,x) \cdot e^{-r \cdot t} dt$

Subject to: $\dot{x} = G(x) - q - y,$
$\quad\quad\quad\quad q, y \geq 0,$

where, it may be recalled, q represents harvest of the commercial fishers and y that of the recreational fishers.

The necessary conditions for a nontrivial solution ($q>0$) to this problem (Pontryagin et al. 1962) lead to the following equations:

(I.1) $\quad y = 0,$

(I.2) $\quad \Pi_{qq} \cdot \dot{q} + \Pi_{qx} \cdot \dot{x} = -\Pi_x + (r - G_x) \cdot \Pi_q$

(I.3) $\quad \dot{x} = G(x) - q.$

In equilibrium, $\dot{q} = \dot{x} = 0$. Therefore, in equilibrium, the above conditions are reduced to:

(4) $\quad \begin{array}{l} y = 0 \\ G_x + \Pi_x / \Pi_q = r, \\ G(x) = q. \end{array}$

Recreational Fishers

Recreational fishers seek to solve the maximization problem:

(II) $\quad \underset{q,y}{Max} \int_0^\infty U(q,x) \cdot e^{-r \cdot t} dt$

Subject to: $\quad \dot{x} = G(x) - q - y.$
$\quad\quad\quad\quad q, y \geq 0$

The necessary conditions for a nontrivial solution ($y>0$) to this problem (Pontryagin et al. 1962) lead to the following equations:

(II.1) $\quad q = 0,$

(II.2) $\quad U_{yy} \cdot \dot{y} + U_{yx} \cdot \dot{x} = -\Pi_x + U_y \cdot (r - G_x),$

(II.3) $\dot{x} = G(x) - q$,

In equilibrium, $\dot{y} = \dot{x} = 0$. Therefore, in equilibrium, the above conditions are reduced to:

(5) $\quad \begin{aligned} & q = 0 \\ & G_x + U_x/U_q = r, \\ & G(x) = y. \end{aligned}$

The above solutions to the maximization problems of the two parties are formally identical. This is not surprising. Both activities are qualitatively the same, extraction of fish to generate benefits. The difference between the two sets of optimality conditions is entirely due to the different objective functions.

As a consequence of the two different objective functions, the two parties will in general want different fisheries policies. As demonstrated by the two sets of solution conditions above, this applies both in equilibrium and disequilibrium, that is, along the optimal dynamic paths. Perhaps, most strikingly, each party would always prefer the extraction rate of the other party to be zero (I.1 and II.1). Moreover, they would in general want a different equilibrium biomass as indicated by the equilibrium conditions, (4) and (5) above. Hence their ideal aggregate extraction rate for each level of biomass would generally also differ. In other words, there is a fundamental conflict between the two parties desired use of the resource. Figure 5.2 illustrates an example of the optimal fisheries policy for the two parties.

Expression (4) for the commercial fishers is well known in the fisheries economics literature as the equilibrium condition for the optimal fishery (Clark and Munro 1975; Arnason 1990). The term Π_x/Π_q for the commercial fishers in equation (4), which will play some role below, is referred to as the marginal stock effect by Clark and Munro (1975). Given profit maximization, the marginal stock effect is nonnegative and its effect is to increase the equilibrium biomass compared to what would otherwise be the case.

For the recreational fishers, the marginal stock effect as defined in equation (5) is U_x/U_q. This would generally be different from that of the commercial fishers. For instance, if biomass is not so crucial to recreational fishers, the marginal stock effect could be lower.

Individual Transferable Quotas and Commercial Fishing

Individual transferable quotas have become widely and apparently successfully employed in the world's fisheries. According to a recent count (Arnason 2005),

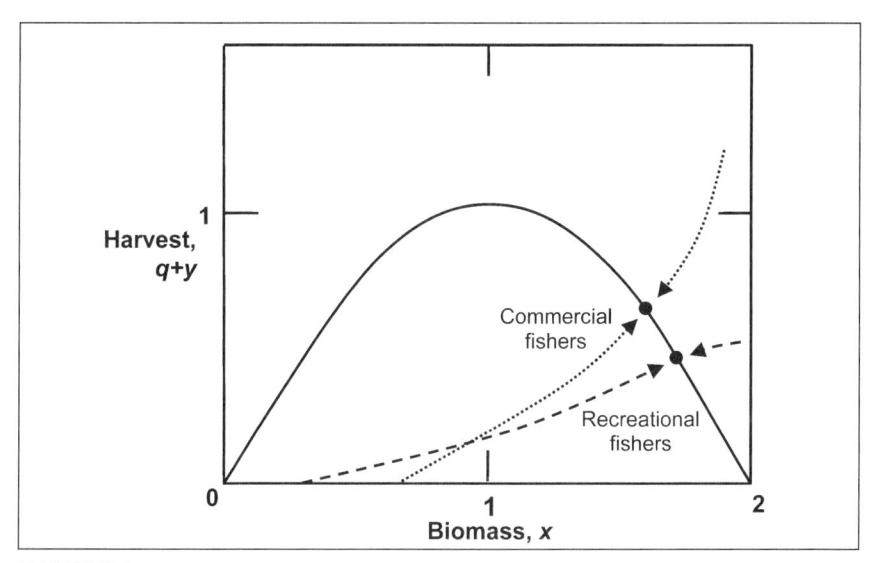

FIGURE 5.2
Desired Biomass and Harvest

at least ten major fishing nations use ITQs as the main or a major component of their fisheries management system and between 10 and 15 percent of the global ocean catch is currently taken under ITQs. Although several other forms of fisheries property rights, such as sole ownership, TURFs, community rights, and so on, exist, ITQs and their non-tradable variant, IQs, seem to be the most widely used property rights instruments in the world's fisheries today.

Let us consider the following ITQ system: There is a total allowable catch, TAC, which applies at each point of time. Individual fishers hold permanent rights to a certain fraction of the TAC. These rights, or quota shares, are perfect property rights in the harvest; that is, they are fully exclusive, secure, permanent, and tradable. It is important to realize, however, that they are fairly weak property rights in the resource itself, namely, the fish stocks and their ocean environment (Arnason 2000).

This ITQ system works by eliminating (or, more precisely, neutralizing) the stock externality which is the predominant cause of the fisheries problem. It does so in two ways: First, by setting the TAC the total extraction level is fixed.[6] Therefore, the evolution of the stock over time becomes exogenous to the harvesting decisions of the fishers.[7] Second, due to the individual quota constraint, no fisher can impose externalities on the other fishers, at least not above what his quota permits. A fisher cannot increase his share of the TAC unless other fishers agree to give him some of their quota rights. Thus, in a fundamental way, the stock externality, the most damaging externality in the fishery, is eliminated (Arnason 1990). Note that the externality disappears by virtue of the property rights created by the ITQs. This illustrates the basic theorem[8] that externalities always arise as a consequence of missing or imperfect property rights. It immediately follows that if the property rights value of the ITQs is somehow reduced, for instance by imperfect enforcement, the stock externality will generally reappear.

ITQs are property rights in harvest. Provided the ITQs are fully tradable, it follows that the allocation of stock use, (i.e., harvests) will be economically efficient (Arnason 1990). Only the most efficient fishing firms will harvest the TAC. The reason for this is simple. If a less efficient firm holds an ITQ, it will find it profitable to sell or rent a part of it to a more efficient fishing firm. This applies to all firms at all times. Thus, as a result, at each point of time, only the most efficient firms will be engaged in harvesting. Moreover, the allocation of harvests between active firms will be economically efficient. For that to be the case, they must all be operating at a point where they generate the same marginal benefits. This is precisely what happens under the ITQ system. Again the reason is profit maximizing trading. If one firm is operating at a lower marginal benefit than another, they will both profit from transferring a part of the first firm's ITQs to the second. Hence, in a trading equilibrium, the marginal benefits of quota use will be equal. This common marginal benefit is also the equilibrium price at which quotas are exchanged.

Thus, we see that under this ITQ system, the right number of the most efficient fishing firms or fishers operating at the optimal level will do the harvesting. In other words, the harvest will be allocated efficiently. A very important thing to note is that this allocative efficiency happens in a totally decentralized manner. It is brought about by individual fishers trying to maximize their own benefits. This they can do by reallocating quotas amongst themselves through trading. These trades, for the usual economic reason, are in the direction of more efficient allocation. So, allocative efficiency of the ITQ system depends critically on the tradability of the quotas. Interestingly, however, it does not depend on how the quota shares are initially allocated. Irrespective of the initial allocation of quota shares, profit maximizing trading will always move them to the most efficient fishing firms.

We can illustrate this basic allocative efficiency of the ITQ system with a simple diagrammatic device. Consider the allocation of harvests between any two fishers. For analytical purposes, we do not have to consider all fishers at the same time because one of them, fisher 2, say, could represent all the remaining fishers. Having determined the allocation of harvests between fisher 1 and the rest, we could move on to the first fisher in the remaining group and so on. Now, each fisher receives certain marginal benefits from harvesting. Let us refer to these two marginal benefit functions as $\Pi_q(q(1),x)$ and

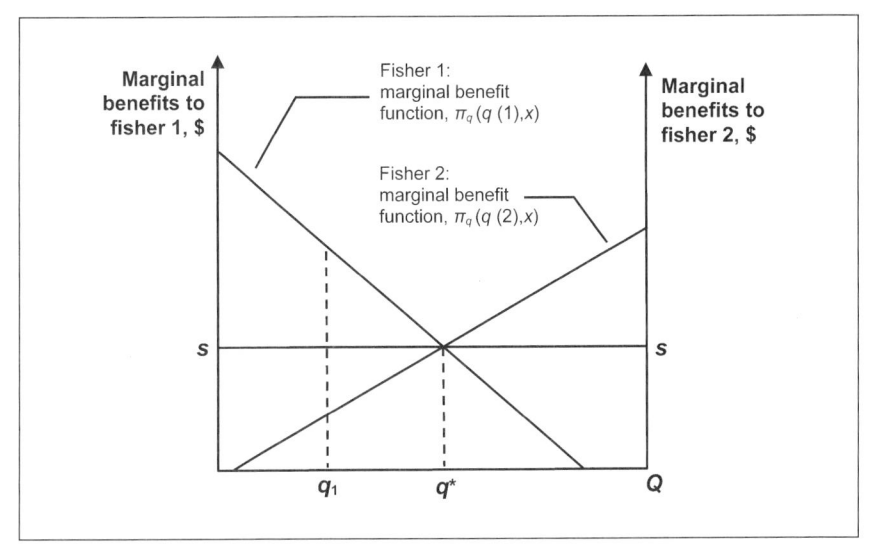

FIGURE 5.3
Optimal Allocation of Resource Use (harvests)

$\Pi_q(q(2),x)$, respectively and draw them as in figure 5.3. In this figure, the marginal benefits to fisher 1 are measured on the left-hand vertical axis and the marginal benefits to fisher 2 on the right-hand vertical axis. The total harvest to be allocated (i.e., the TAC) is measured along the horizontal axis between the two vertical axes and we refer to it as Q. Any point on the horizontal axis represents a given allocation of harvests between the two fishers. Thus, for instance, the point q_1 on the horizontal axis represents the allocation of q_1 units to fisher 1 and the remaining $(Q-q_1)$ units to fisher 2.

Now, it is easy to establish that the economically optimal allocation of the total harvesting quantity, Q, between the two fishers occurs at the point q^*, where their two marginal benefit curves intersect. For instance, letting fisher 1 harvest a little bit more and fisher 2 a little bit less (i.e., moving slightly to the right of q^*) entails less gains to fisher 1 than there are losses to fisher 2. Thus, this modification cannot be economical. Corresponding arguments apply to letting fisher 2 harvest a little bit more and fisher 1 a little bit less.

It is similarly easy to establish that under the ITQ system, q^* is precisely the point to which the allocation of harvests between the two fishers will converge, irrespective of the initial allocation of quotas. To see this, we only have to note that whenever they are not at q^*, they will both benefit from trading their quotas toward q^*. Thus, we can use this diagrammatic device to see that the ITQ system results in efficient allocation of the total allowable catch (TAC) between fishers.

At the optimal allocation point, q^*, the two marginal benefit curves are equal at the value s in the diagram. As already stated, this will be the equilibrium quota price in the quota market. It may be interesting to note that the area $s \cdot Q$ in figure 5.3 is a measure of the resource or fisheries rents in this fishery.

Note that there is no reason for the optimal point q^* to correspond to positive harvests for both fishers. If one of them (e.g., fisher 1) is very inefficient relative to the other, his marginal benefit curve will be very low and fisher 2 will take all the allowable harvest

Thus, for any TAC setting, the fishery will operate as efficiently as possible. This strong theoretical result seems to be verified by the experience of ITQ systems around the world (Wilen and Homans 1994; Shotton 2000; Arnason 2005). The problem, however, is to select the right TAC. This basically has to be set in such a way that the fishery follows the optimal path toward equilibrium. It is straightforward to define this path analytically. Due to the lack of necessary information, however, to find it in practice is a different proposition. Provided markets work, it can be shown that the optimal TAC at each point of time is the one that maximizes the permanent quota values (e.g., Arnason 1990). Thus the problem of finding the right TAC is reduced to selecting the one that maximizes the value of quota shares. Fortunately, it turns out in most empirical cases that this value function is quite flat around its maximum level. Thus, usually, it is not of any great consequence to make relatively small errors in setting the annual TACs provided the errors are not one-sided; that is, consistently above or below the optimal TAC.

Typically, in most countries that employ ITQs, the TACs are set by the fisheries authorities. However, there is no reason for that to be so. Any agency that can locate the maximum of the share quota value function is capable of setting the optimal TAC. In fact, due to their overriding incentive to maximize their wealth and their intimate knowledge of the fishery, it appears that the industry, acting as a whole, is probably in the best position to set TAC itself (Arnason 2007).

Reconciling Commercial and Recreational Fishing under ITQs

We have seen how ITQs can harmonize conflicting demands for commercial harvests from fish stocks. What about other extraction demands from, say, recreational fishers? The answer is that these demands can be completely reconciled with commercial fishing demands within the framework of the ITQ system. The reason is simple. Recreational demand for harvests is analytically identical to the commercial one. Recreational fishing is extractive in exactly the same way as commercial fishing, and recreational fishers have benefit functions qualitatively the same as those of the commercial fishers. As a result, recreational fishers are, for analytical purposes, just like additional commercial fishers. They should harvest to the extent that their marginal benefit functions exceed those of the commercial fishers already active in the fishery. At the same time, for any given TAC, less beneficial recreational fishing should give way to make room for the commercial fishing.

The social objective is to maximize the present value of total benefits. In other words:

(III) $\underset{q,y}{\text{Max}} \int_0^\infty [\Pi(q,x) + U(y,x)] \cdot e^{-rt} dt$

Subject to: $\dot{x} = G(x) - q - y,$

$q, y \geq 0.$

where, as before, $\Pi(q,x)$ is the benefit function of the commercial fishing industry and $U(y,x)$ is the benefit function of the recreational fishery. The variables q and y denote the harvest of the commercial fishery and the recreational fishery, respectively, the sum of which represents the TAC.

Necessary conditions to solve this problem include the conditions:

$\Pi_q = \lambda = U_y$, for all active fishers.
$\dot{\lambda} - r \cdot \lambda = -\Pi_x - U_x - \lambda \cdot G_x.$
$\dot{x} = G(x) - q - y.$

The second and third conditions jointly determine the optimal total harvest at each point of time and the corresponding shadow value of biomass. The first condition, however, is the focus of our investigation. It requires both commercial and recreational fishers (assuming both should be active) to operate where their marginal benefits from harvesting are equal. This, of course, is the key requirement of allocative efficiency in economics.

Under the ITQ system, this is exactly what will be the outcome. If a recreational fisher gets more marginal benefits from fishing than a commercial fisher, it will be in the interest of both to trade some commercial quotas to the recreational fisher and vice versa. This is illustrated in figure 5.4. In this figure, fisher 1 is a recreational fisher and fisher 2 the remaining group of all fishers, commercial and recreational. The optimal allocation of harvests is q^*, which, as argued above, is going to be brought about by quota trades. At this optimal point the marginal benefits to both the recreational fisher in question and the other fishers are the same and equal to the market price for quotas, s. This market price, however, would normally differ from the one applying

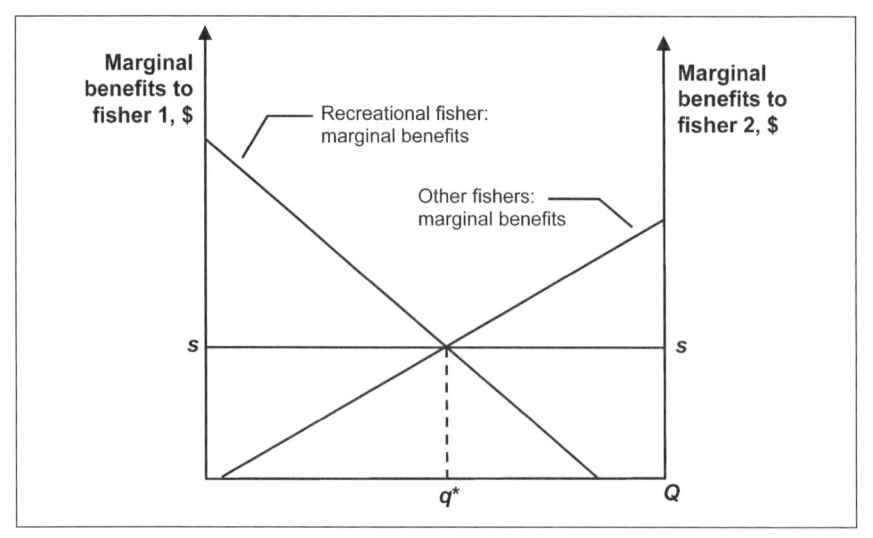

FIGURE 5.4
Commercial and Recreational Marginal Benefits

if only the commercial fishers were included in the quota system. In fact, it would normally be higher. Note, moreover, that as before, it doesn't matter for allocative efficiency how the quotas are initially allocated. Quota trades will bring them to the most efficient point. Finally, note that, as before, the optimal allocation does not necessarily have to imply that both sectors, the commercial and the recreational, are active in the industry. If one of them is sufficiently efficient, the other could simply be eliminated from active participation in the fishery through trades.

The above discussion ignores the setting of the optimal TAC level. With the addition of the recreational sector, the optimal TAC, that is, the one that maximizes the value of the total fishery (recreational and commercial), would generally be changed. More specifically, with recreational fishing included, the optimal equilibrium conditions for the fishery will be defined by the conditions:

(6) $$\begin{aligned} G_x + (\Pi_x + U_x)/\Pi_q &= r, \\ G(x) &= q+y. \\ \Pi_q &= U_y. \end{aligned}$$

As before, $\Pi(q,x)$ is the benefit function of the commercial fishing industry and $U(y,x)$ the benefit function of the recreational fishery. The variables q and y denote the harvest of the commercial fishery and the recreational fishery, respectively, the sum of which represents the optimal TAC.

Comparing this new set of conditions to the ones for the commercial fishery only, expression (4) above shows that the presence of the recreational fishery alters the marginal stock effect—it is now $(\Pi_x+U_x)/\Pi_q$ instead of Π_x/Π_q. So, the new marginal stock effect takes account of the marginal benefits of biomass to the recreational fishers as well to the commercial fishers. This means that in general both the optimal biomass and the corresponding TAC will be altered. If both industries should operate, the jointly optimal biomass will be some average of the ones the parties would have selected for themselves. Normally, if the recreational fishers value the stock more highly than the commercial fishers, the TAC will be reduced compared to what the commercial fishers would like and vice versa. The rule for setting the optimal TAC is basically unchanged. It should be set so as to maximize the total value of all quota shares, recreational and commercial.

Separate Property Rights Systems

We now briefly consider the situation where there are two separate property rights systems in the resource. For convenience, let us proceed as if these property rights systems are ITQs. ITQ-systems may be seen as examples of other property rights systems of a similar property rights quality (Arnason 2005). Therefore, the outcome for ITQ systems should also be indicative of the outcome under other equally high quality property rights systems.

We specify the following situation: the biomass evolves according to the following equation:

$$\dot{x} = G(x) - Q,$$

where Q is the total TAC. This, of course, is basically as before in equation (1).

The total TAC is now divided into the commercial TAC and the recreational TAC, R and T, respectively.

$$Q = R + T.$$

Finally, let us assume that the commercial fishers hold property rights in the commercial system and the recreational fishers in the recreational system.

Nonintegrated Systems

First assume that the two property rights systems are nonintegrated in the sense that trading of rights from one group to another is not possible. Then it is easy to show (Arnason 1990 and Appendix 5.1) that there will be two different prices of quota shares (more generally the property rights) in the two systems. Consequently, the marginal benefits in the two systems will differ and the resource use will be inefficient. More precisely:

(7) $\Pi_q(q,x) = \dfrac{r \cdot s - \dot{s}}{R}$,

(8) $U_y(y,x) = \dfrac{r \cdot w - \dot{w}}{T}$,

where s and w refer to the property rights prices in the commercial and recreational systems, respectively. Note that these prices refer to the total property right and the division by R and T serves to obtain prices per unit of harvest. So the right hand sides of (7) and (8), respectively, are really the momentary holding costs of the two property rights.

It follows immediately from (7) and (8) that the marginal products will not be identical except by coincidence and therefore the resource use will be inefficient.

Integrated Systems

Now, consider the situation where the two property rights systems are integrated in the sense that trading of rights from one group to another is possible (and costless). In that situation commercial fishers will generally trade quotas with recreational fishers and price equalization over the two assets will ensue. As a result, at least in asset market equilibrium, the marginal benefits of recreational and commercial fishers will be equal and the resource use will be fully efficient.

More formally we can show (see also Appendix 5.2):

(9) $\Pi_q(q,x) = \dfrac{r \cdot s - \dot{s}}{R} = U_y(y,x)$,

(10) $U_y(y,x) = \dfrac{r \cdot w - \dot{w}}{T} = \Pi_q(q,x)$.

where, as before, s and w refer to the property rights prices in the commercial and recreational systems, respectively.

The key message of (9) and (10) is that there will be allocative efficiency between recreational and commercial fishing under an integrated ITQ system (or another equally high quality property rights system). Note that this allocative efficiency occurs spontaneously as a result of mutually advantageous trading. Note, moreover, that it is independent of the respective TAC allocations to the systems. Thus, provided that such(costless) trading between the two property rights systems is possible, it doesn't matter for efficiency how the initial property rights are assigned.

Conclusion

Limited natural resources give rise to resource conflicts. In the case of a fish stock whose use is shared by fishers, competition for harvests and, therefore, conflicts can occur both within a particular type of use and between uses (e.g., within a sector and between sectors). This chapter is concerned with the competition for harvests and, therefore, conflicts between commercial fishers, on the one hand, and recreational fishers, on the other.

It was shown that a properly designed and operated ITQ system is capable of resolving conflicts in resource use among commercial fishers in the socially optimal way, conditional, of course, on the TAC that has been set. Including recreational fishing in the same ITQ system will produce the socially optimal allocation of resource use across all fishers, commercial and recreational. Given the importance of recreational fishing in many fisheries, this is potentially a very useful result. Note, however, that this is derived assuming perfect and costless enforcement of the ITQ system. Obviously, costly and imperfect enforcement might lead to certain modifications of the result. In certain fisheries, it may, for instance, turn out that the enforcement of ITQ restrictions on recreational fishers is prohibitively costly.

The reason why the ITQ system so easily harmonizes the interests of both commercial and recreational fisheries is that the two uses of the resource are, from an economic perspective, essentially the same—both extract harvests from the resource—and the good (i.e., the harvest) they obtain is their private property. As a result, one price, the quota price, is sufficient to coordinate their interests in the socially optimal manner. Of course, the problem of setting the overall quota or total allowable catch remains. However, as discussed elsewhere, procedures for doing so effectively exist (Arnason 1990).

Harmonizing conservation interests in a fish stock with those of commercial and recreational interests, however, is not as straightforward. The basic reason is that the benefits to conservationists, presumably a larger stock, is not a private, but a public, good. Not only is it a public good to the conservationists, it is also a public good (a common benefit) to commercial and recreational users. When it comes to conservation interests, the key variable is the TAC. Obviously, a decentralized trading of catch quotas based on an already set TAC cannot determine the correct TAC. For that task further modifications of the standard ITQ system are required.

These basic results seem to generalize to other property rights systems in fisheries, provided the property rights quality is not less than that of an ITQ system. The results also generalize to two distinct ITQ or other property rights systems in the same resource, provided the systems are integrated in the sense that trading of property rights between them is possible.

Notes

1. A Pareto-efficient solution is reached when no party can be made better off without making another party worse off.

2. Interesting examples of these social institutions are those that governed fishing access to Pacific salmon in North America prior to white settlement. See, for example, Higgs (1982) and Nikel-Zueger (2003).

3. In some cases, especially the United States and Canada, the state claims ownership of salmon rivers.

4. In Scotland, the shift from commercial to recreational fishing of Atlantic salmon that started in the latter half of the nineteenth century was facilitated by trade between upriver angling interests and holders of netting rights downstream. See Robertson (1988).

5. Actually, by reference to basic market principles, if financial markets are perfect this should be the case.

6. This obviously assumes that individual quota constraints are enforced.

7. It has been shown by Arnason (1990) that if the quota constraint is binding (i.e., the ITQs impose restrictions on some fishers) it is never optimal for commercial fishers to leave quota unused.

8. In my personal notes I have proven this theorem to my satisfaction. Interestingly, although the theorem is important and apparently widely thought to be true, I have not come across a formal statement of it, let alone proof, in the literature.

References

Arnason, Ragnar. 1990. Minimum Information Management in Fisheries. *Canadian Journal of Economics* 23: 630–53.

———. 2000. Property Rights as a Means of Economic Organization. In *Use of Property Rights in Fisheries Management*, ed. Ross Shotton. FAO Fisheries Technical Paper 404/1, 14–25. Proceedings of the FishRights99 Conference Fremantle, Western Australia, November 11–19, 1999.

———. 2005. Property Rights in Fisheries: Iceland's Experience with ITQs. *Reviews in Fish Biology and Fisheries* 15: 243–64.

———. 2007. Fisheries Self-management under ITQs. *Marine Resource Economics* 22: 373–90.

Clark, Colin W. 1976. *Mathematical Bioeconomics: The Optimal Management of Renewable Resources.* New York: John Wiley & Sons.

Clark, Colin W., and Gordon R. Munro. 1975. The Economics of Fishing and Modern Capital Theory: A Simplified Approach. *Journal of Environmental Economics and Management* 2: 92–106.

Gordon, H. Scott. 1954. Economic Theory of a Common Property Resource: The Fishery. *Journal of Political Economy* 62: 124–42.

Hatcher, Aaron, Sean Pascoe, Richard Banks, and Ragnar Arnason. 2002. *Future Options for UK Fish Quota Management: A Report to the Department for the Environment, Food and Rural Affairs.* CEMARE Report 58. Portsmouth, UK: University of Portsmouth, Centre for the Economics and Management of Aquatic Resources, June.

Higgs, Robert. 1982. Legally Induced Technical Regress in the Washington Salmon Fishery. *Research in Economic History* 7: 55–86.

International Council for the Exploration of the Sea (ICES). 2007. *Report of the Working Group on Working Group on North Atlantic Salmon (WGNAS): ICES WGNAS Report 2007.* ICES CM 2007/ACFM:13. Copenhagen: ICES.

Lackey, Robert T. 2005. Fisheries: History, Science and Management. In *Water Encyclopedia: Surface and Agricultural Water,* ed. Jay H. Lehr and Jack Keely. New York: John Wiley & Sons, 121–29.

National Research Council (NRC). Committee to Review Individual Fishing Quotas. 1999. *Sharing the Fish: Toward a National Policy on Individual Fishing Quotas.* Washington, DC: National Academies Press.

Nikel-Zueger, Manuel. 2003. Saving Salmon the American Indian Way. *PERC Policy Series,* PS-29. Bozeman, MT: PERC.

Organization for Economic Cooperation and Development (OECD). 1997. *Toward Sustainable Fisheries: Economic Aspects of the Management of Living Marine Resources.* Paris: OECD.

Pontryagin, Lev S., Vladimir S. Boltyanski, Revaz V. Gamkrelidze, and Evgenii F. Mishchenko. 1962. *The Mathematical Theory of Optimal Processes.* New York: John Wiley & Sons.

Robertson, Iain A. 1998. *The Tay Salmon Fisheries Since the Eighteenth Century.* Glasgow, UK: Cruithne Press.

Shotton, Ross, ed. 2000. *Use of Property Rights in Fisheries Management.* Proceedings of the FishRights99 Conference. Fremantle, Western Australia, November 11–19, 1999. *FAO Fisheries Technical Paper* 404/1 and 404/2. Rome: Food and Agriculture Organization of the United Nations.

Smith, Adam. 1977. *An Inquiry into the Nature and Causes of the Wealth of Nations,* ed, Edwin Cannan. Chicago, IL: University of Chicago Press.

Wilen, James E., and Francis R. Homans. 1994. Marketing Losses in Regulated Open Access Fisheries. In *Fisheries Economics and Trade: Proceedings of the Sixth Conference of the International Institute of Fisheries Economics and Trade,* ed. Joseph Catanzano. Paris: IFREMER, 795–801.

Appendix 5.1

Several Nonidentical Commercial and Recreational Fishers

Consider i commercial fishers and j recreational fishers. Let the benefit function of the former be:

$$\Pi(q(i), x; i), \ i=1,2,\ldots I$$

where $q(i)$ represents the harvest of fisher i and biomass x. Similarly, let the benefit function of recreational fisher j be:

$$U(y(j), x; j), \ j=1,2,\ldots J$$

The social objective is to maximize the present value of total benefits. In other words:

$$\underset{\forall q(i), y(j)}{Max} \int_0^\infty \sum_{i=1}^I \Pi(q(i), x; i) + \sum_{j=1}^J \Pi(y(j), x; j) \cdot e^{-rt} dt$$

Subject to: $\dot{x} = G(x) - \sum_{i=1}^I q(i) - \sum_{j=1}^J y(j),$

$$q(i), y(j) \geq 0, \ \forall i, j.$$

Necessary conditions to solve this problem include the conditions:

(A.1) $\Pi_{q(i)} = \lambda = U_{y(j)}$, for all active fishers.

$$\dot{\lambda} - \rho \cdot \lambda = -\sum_{i=1}^{I} \Pi_\xi - \lambda \cdot \Gamma_\xi.$$

$$\dot{x} = G(x) - \sum_{i=1}^{I} q(i) - \sum_{j=1}^{J} y(j).$$

The first condition requires all active fishers (the inactive ones are not excluded from the fishery) to operate where all marginal profits of active ($q(i)>0$) commercial fishers from harvesting are equal to all marginal benefits of active ($y(i)>0$) recreational fishers. This is the key requirement of allocative efficiency. The second and third conditions jointly determine the optimal total harvest at each point of time and the corresponding shadow value of biomass.

Now consider an integrated ITQ system. Under this system, any arbitrary commercial fisher i holds a certain share $\alpha(i)$ in the total allowable catch, Q. Similarly, any arbitrary recreational fisher j holds a certain share $\alpha(j)$ in the total allowable catch, Q. All fishers can buy and sell quota shares (and quotas) at the going market price s. Any arbitrary commercial fisher thus attempts to solve the following problem:

$$\underset{q(i)}{Max} \int_0^\infty (\Pi(q(i), x; i) - s \cdot z) \cdot e^{-r \cdot t} dt$$

Subject to: $q(i) \leq \alpha(i) \cdot Q$
$\dot{\alpha}(i) = z$
$\dot{x} = G(x) - Q,$
$q(i) \geq 0.$

Note that as regards solving the maximization problem, the second constraint is redundant as it contains no variables the fisher can influence.

Necessary conditions to solve this problem for all active fishers include the conditions:

(A.2) $\Pi_{q(i)} = \dfrac{r \cdot s - \dot{s}}{Q}, \forall i.$

It immediately follows that $\Pi_{q(i)} = \Pi_{q(j)}$ for all active fishers. This is equivalent to the requirement of allocative efficiency in the social optimal problem. This proves that the ITQ system generates allocative efficiency at each point of time. If, moreover, the $\dfrac{r \cdot s - \dot{s}}{Q} = \lambda$, the fishery will also follow the optimal biomass path over time. This will happen if the TAC (i.e., Q) is set correctly at each point of time.

A corresponding profit maximization exercise for active recreational fishers leads to the conditions:

(A.3) $U_{y(j)} = \dfrac{r \cdot s - \dot{s}}{Q}$, $\forall j.$

Combining (A.2) and (A.3) yields the social optimality requirement, (A.1).

Appendix 5.2
Two Distinct ITQ Systems

Let the total quota Q be divided into a commercial and recreational quota:

$Q = R + T$.

Let the market prices of the recreational and commercial quotas be s and w, respectively.
Let the two groups of fishers' share of the quotas be:

$\alpha(l)$ and $\beta(l)$, $l = 1, 2$.

And let their net purchases of the quotas be:

$z(l)$ and $u(l)$, $l = 1, 2$.

The profit maximization problem for the commercial fishers is:

$$\underset{z(1), u(1)}{Max} \int_0^\infty (\Pi(\alpha(1) \cdot R + \beta(1) \cdot T, x; i) - s \cdot z(1) - w \cdot u(1)) \cdot e^{-rt} dt$$

subject to: $\dot{\alpha}(1) = z(1)$
$\dot{\beta}(1) = u(1)$
$\dot{x} = G(x) - Q,$
$1 \geq \alpha(1), \beta(1) \geq 0$.

And the benefit maximization problem for the recreational fishers is:

— 119 —

$$\underset{z(2),u(2)}{\text{Max}} \int_0^\infty (U(\alpha(2)\cdot R + \beta(2)\cdot T, x; i) - s\cdot z(2) - w\cdot u(2))\cdot e^{-rt} dt$$

subject to: $\dot{\alpha}(2) = z(2)$
$\dot{\beta}(2) = u(2)$
$\dot{x} = G(x) - Q,$
$1 \geq \alpha(2), \beta(2) \geq 0$.

Solving these two profit maximization problems (assuming both sectors are active) leads to the following optimality conditions:

$$\Pi_q(q,x) = \frac{r\cdot s - \dot{s}}{R} = U_y(y,x),$$

$$U_y(y,x) = \frac{r\cdot w - \dot{w}}{T} = \Pi_q(q,x).$$

If there is more than one (active) fisher in each group, these same conditions apply to each one of them.

Part III
IFQS AND THE COMMERCIAL CHARTER BOAT SECTOR

Chapter 6

Examining the Interface between Commercial Fishing and Sportfishing: A Property Rights Perspective

Keith R. Criddle

Recommendations favoring adoption of rights-based management strategies have been developed primarily in the context of commercial fisheries and have focused on increasing the profitability of commercial fishing and reducing the incentive to deplete fish stocks. Relatively little attention has been given to the effects that alternative management regimes could have on the profitability of processing and support service businesses, on the magnitude of net benefits to consumers, or to the interface between commercial fishing, sportfishing, and other use and nonuse demands for fishery resources. Lack of concern about changes in the magnitude of net benefits to buyers and consumers of commercially harvested fish can be attributed to the implicit assumption that there are readily available substitutes for fish harvested in every particular fishery, an assumption that is not supported in empirical analyses (see, for example, Cheng and Capps 1988; Lin, Richards, and Terry 1988; Roy, Tsoa, and Schrank 1991; Herrmann, Mittelhammer, and Lin 1993; Eales, Durham, and Wessells 1997; Salvanes and DeVoretz 1997; Herrmann and Criddle 2006). It is similarly unreasonable to focus exclusively on the commercial fishery in as much as most fish stocks support a variety of commercial, sport, and other use and nonuse services, the provision of which may be subject to conflicting management regimes and the benefits of which accrue to different recipients.

This chapter presents a simple conceptual analysis of the effects of alternative regimes for management of a charter-based sport fishery[1] on the

---

This chapter updates Criddle (2004).

magnitude of net benefits in sport and commercial fisheries. An empirically based comparative static simulation is used to identify the optimal commercial–sport allocation and the optimal sustainable yield for the commercial and charter-based sport fisheries for halibut off Alaska.

A Conceptual Analysis of the Commercial Fishing and Sportfishing Interface

Four stylized management regimes for commercial and charter-based sportfishing are examined in this section. For simplification, it is assumed that the target species is not subject to other uses, such as subsistence, self-guided sportfishing, or bycatch,[2] that nonuse values are inconsequential, that trophic interactions with other species are insignificant, and that the fishery is governed by a management authority that establishes and enforces a sustainable binding overall cap on the sum of commercial and sport catches. In addition, it will be assumed that landed catches in the sport fishery are a constant multiple of the number of sportfishing trips taken and that the nonpecuniary attributes of sportfishing trips are immutable. That is, sportfishing charters are assumed to compete through price alone while trip length and amenities are assumed identical across charter services. Relaxing these simplifying assumptions would complicate description of the outcomes without materially changing the conclusions.

Regulated Open-Access Commercial Fishing and Open-Access Sportfishing

The most common combination of commercial and sport fisheries involves a commercial fishery subject to an overall catch quota but lacking efficacious constraints on fishing capacity and a sport fishery that lacks binding constraints on overall catch or fishing capacity. While the economic consequences of regulated open-access management of the commercial fishery are explored in, for example, Homans and Wilen (1997) and Wilen and Homans (1998), the interaction of a regulated open-access commercial fishery and an open-access sport fishery has not been examined in depth.

Anticipated increases in the demand for sportfishing trips are usually accommodated through reductions in the overall commercial quota. Similarly, reductions in stock abundance that lead to reductions in the overall cap of sport and commercial harvests typically lead to reductions in the overall commercial quota. Changes in the exvessel (dockside) demand for fish caught in the commercial fishery will be assumed to affect the exvessel price but not to influence the allocation between sport and commercial fisheries. Because the number of sportfishing charters is large and because barriers to entry are relatively small, it can be assumed that sportfishing charters behave as perfect competitors. That is, if individual charter operators were to attempt to raise trip prices, they would lose customers to competitors who did not raise trip prices. In addition, because of the low barriers to entry, the capacity of the charter fleet will quickly expand or contract in response to changes in the demand for sportfishing trips.

Given these assumptions, an increase in the demand for sportfishing trips will lead to an increase in the number of sportfishing trips taken, an increase in the total value to anglers beyond the amount they pay for trips, and an increase in charter capacity. Although sportfishing trip prices and charter operator profits may rise in the short run, prices will be bid back to their initial levels as the charter fleet expands. In this example, because the commercial fishery is constrained by an overall quota but the number of participants or level of fishing effort is unregulated, the ownership-by-capture rule leads to the adoption of cost-increasing technologies so that in equilibrium the financially marginal commercial harvester earns only enough to cover operating and opportunity costs. In these circumstances, when increased demand for sportfishing trips is accommodated by a reduction of the overall commercial quota, the immediate effects in the commercial fishery will be an increase in exvessel prices and a reduction in net benefits to the buyers and consumers of commercially harvested fish. The short-run change in profits to commercial harvesters could be positive or negative depending on the availability of substitute seafoods and the magnitude of the reduction in the commercial quota. In either case, reduction of the commercial quota will lead to an intensified race for catch. Thus in the long run, increased sportfishing demand will lead to increased exvessel price, reduced commercial profits, and reduced net benefits to buyers and consumers of commercially harvested fish.

Similarly, modest fluctuations in abundance of the target species that lead to small changes in the allowed sum of commercial and sport catches will be reflected in the overall commercial quota while leaving the sportfishing allocation unaffected. Consequently, modest stock fluctuations would not be likely to result in short-run or long-run changes in the earnings of sportfishing charters or in the total value to anglers beyond the amount they paid for sportfishing trips. Because fluctuations in the abundance of target species are generally reflected in the overall commercial quota, fluctuations in abundance will result in short-run profits or losses in the commercial fishery. In the long run, the lack of effective constraints on effort and the dynamics of regulated open access will dissipate any short-run profits while losses will lead to business failures and attrition in the number of participants in the commercial fishery. Abundance increases (decreases) can be expected to increase (decrease) net benefits to fish buyers and consumers.

Because changes in the exvessel demand affect exvessel price but do not influence the allocation between sport and commercial fisheries, the effect of such changes will be similar to the effect of fluctuations in abundance. That is, if the exvessel demand increases, the resultant short-run profits will attract increased effort that will ultimately dissipate the short-run profits in the commercial fishery.

In each of these cases, whether the allocation of a larger share of the catch to the charter-based sport fishery will result in a net gain to society depends on whether the long-run increase in total value to anglers beyond the amount they pay for trips exceeds the long-run decrease in benefits that fish buyers and consumers derive from catches sold by the commercial fishery. That is, when there is open access in the commercial and charter fisheries, growth in the sport fishery represents a transfer of benefits from buyers and consumers of commercially harvested fish to anglers with a contraction of the commercial fishing sector and an expansion of capacity in the sportfishing charter sector.

Individual Quota-Based Commercial Fishing and Open-Access Sportfishing

Economists have long argued that a well-designed individual fishing quota (IFQ) program can increase the profitability of commercial fishing and possibly reduce the incentive to overharvest fish stocks. However, little attention has been given to the interface between an IFQ-managed commercial fishery and other fisheries that exploit the same fish stock. In this and subsequent examples, it will be assumed that the commercial fishery is subject to an IFQ regime wherein individual harvesters hold exclusive rights to harvest a percentage of an overall quota set by the management authority. It will also be assumed that the charter-based sport fishery lacks efficacious constraints on fishing capacity or total sportfishing catches and that the management authority accommodates anticipated increases in the demand for sportfishing trips through reductions in the overall commercial quota that lead to proportionate reductions in the annual realization of the IFQs. This latter assumption will be relaxed in subsequent examples. Because assumptions about the sport fishery are unchanged from the first example, the short- and long-run effects on the sport fishery of an increase in the demand for sportfishing trips, moderate changes in halibut stock abundance, or changes in the exvessel demand for commercial catches are also unchanged. But because the commercial fishery is IFQ-based, the effects on the commercial fishery are markedly different from those in the preceding example.

In addition to assuming that growth in the demand for sportfishing will be accommodated through reductions in the overall commercial quota, it will be assumed that reductions in the overall commercial quota are automatically distributed as proportionate reductions in individual IFQs. Given these assumptions, the immediate effects of a reduction of the commercial quota will be an increase in the exvessel price and a decrease in benefits derived by fish buyers and consumers. Depending on the availability of substitute seafoods and the specific quota level, commercial harvesters could be advantaged or disadvantaged by the change in overall commercial quota; however, the sum of commercial profits and net benefits to fish buyers and consumers will be unambiguously reduced. Nevertheless, the loss in commercial fishing profits will be smaller under IFQ management than under regulated open access. That is, IFQ-based management helps the commercial fishery mitigate the adverse impacts of increased allocations to the sport fishery. In the long run, struggling commercial harvesters will choose to sell their IFQs to lower-cost operators; thus an expansion of sportfishing demand can be expected to lead to additional consolidation of IFQ holdings. Another difference from the open-access case is that increased sportfishing demand affects the asset value of the IFQ. That is, reductions in the commercial quota affect the wealth of commercial harvesters as well as their current revenues. Although the change in commercial profits could be positive or negative for small changes in the demand for sportfishing, it will be negative for large increases in the demand for sportfishing. The asset value of the IFQ will depend on expected future profits and will incorporate expectations regarding anticipated changes in the commercial quota associated with future changes in the demand for sportfishing.

Moderate fluctuations in the abundance of the target fish population will lead to changes in the overall commercial quota with inconsequential changes in angler success or in the total catch taken in the sport fishery. Because abundance fluctuations affect the magnitude of commercial catches, they affect exvessel price, commercial profits, and net benefits to fish buyers and consumers. Although net benefits to fish buyers and consumers are directly correlated with commercial catch, rising when catches increase and falling when catches decrease, commercial profits could rise or fall in response to increases or decreases in the overall commercial quota, depending on the specific quota level and the availability of substitute seafoods. If the stock fluctuations are high frequency and low amplitude, they are unlikely to affect the asset value of IFQ holdings. If stock fluctuations are long, large, and predictable, they will affect expectations about future catches and profits and cause fluctuations in the asset value of IFQ holdings.

Given the assumptions in this example, changes in the exvessel demand for commercial harvests will affect exvessel price, commercial profits, net benefits to fish buyers and consumers, and the asset value of IFQ holdings. In contrast with the open-access case, increases in profits will be conserved in the asset value of the IFQ rather than dissipated. Similarly, changes in exvessel demand that lead to reductions in commercial profits (e.g., as a result of increased availability of a substitute good such as farmed halibut) will reduce the value of IFQ holdings. As the asset value of IFQs changes, commercial harvesters may choose to increase or decrease their holdings, thereby changing the mix and number of fishery participants. Reductions in the exvessel demand can be expected to lead to consolidation of quota share holdings, while increases in exvessel demand might lead to an increase in the number of commercial fishers. As in the previous example, whether the allocation of a larger share of catch to the sportfishing charter sector will result in a net gain in benefits to society will depend on whether the long-run increase in total value to anglers beyond the amount they paid for trips exceeds both the long-run decrease in profits in the commercial fishery and the long-run decrease in benefits that fish buyers and consumers derive from catches sold by the commercial fishery.

Individual Quota-Based Commercial Fishing and Regulated Open-Access Sportfishing

In the first two examples, it was assumed that the management authority is incapable of or unwilling to control the total magnitude of sportfishing catches. However, management agencies employ a variety of gear restrictions, bag, possession, and retention limits, size restrictions, and seasonal and diurnal closures that influence the magnitude of sportfishing catches. In this example, it will be assumed that some combination of these instruments can be used to limit total charter-based sportfishing catches.

If the charter fleet is composed of small homogeneous firms, a short-run increase in the demand for sportfishing trips will lead to an increase in trip prices and above-normal profits for charter operators. Because the price increase is a result of expanded demand, total value to anglers beyond the amount paid for trips will be unaffected. The long-run effects will depend on what mechanisms are used to constrain sportfishing catches. For example, if the total allowable sportfishing catch is managed through a fishery closure after the cap has been reached, a race for fish is likely to ensue with a shift of catches to earlier and earlier in the year, dissipating potential benefits. Trip limits, limits on the number of days at sea or number of lines fished, and so on, could increase the cost of providing trips or reduce the net benefits associated with trips, thereby reducing net benefits to anglers or profits to sportfishing charters. Although limited entry programs may slow the rate of overcapitalization and rent dissipation, more than three decades of global experience suggests that because it is nearly impossible to constrain all of the dimensions of effort, potential benefits will eventually be dissipated. Because the charter fleet is assumed to be quota constrained in this example, increases in the demand for charter trips will not affect the commercial fishery unless charter operators and anglers successfully lobby for an increased share of the total allowable catch.

In this example, moderate fluctuations in stock abundance or in the exvessel demand for commercial catch will not affect the total net benefits of sportfishing if the allocation between the commercial and sport fisheries is a fixed quota. If the allocation is percentage based, positive fluctuations in stock abundance will lead to short-term gains to charter operators while negative fluctuations will lead to short-term losses. Changes in exvessel demand will affect commercial harvesters and buyers and consumers of commercially harvested fish exactly as they were affected in the previous example. The effect of moderate variations in stock abundance will also be identical if a fixed cap governs the charter fishery. If the allocation between commercial harvesters and charter operators is percentage based, the downside risk associated with changes in fish abundance will be shared by commercial and charter firms. Imposition of a percentage-based hard cap on charter catches would isolate the charter and commercial sectors from each other except to the degree that they lobby for changes in the cap.

Individual Quota-Based Commercial Fishing and Sportfishing

Allocation of catch among competing user groups is a contentious and perennial issue; the oft-rancorous disputes can place paralyzing demands on management agency time. If by some fortuitous circumstance, the political processes arrive at an optimal allocation, optimality will prove short-lived because it is conditional on constant exvessel prices, factor costs, stock abundance, recreation trip costs, opportunity costs of alternative recreational activities, angler success, and so on. Moreover, if the allocation is accomplished through political processes, the allocation battle will be reprised whenever one user group perceives that its negotiating position has improved. Consequently, it is unlikely that allocations between commercial, sport, and other use and nonuse activities will be definitively settled by any single management decision. Under IFQs, voluntary market transactions replace the regulatory allocation process. While IFQs have been implemented in several commercial fisheries, IFQs have not yet been implemented in a U.S. sport fishery. Issues related to the implementation of IFQs in commercial fisheries and the magnitude and distribution of benefits among initial IFQ recipients, subsequent purchasers of IFQ, processors, consumers, and regional economies remain controversial (NRC 1999).

In addition to raising many of the same issues concerning implementation and distribution of commercial IFQs, sportfishing IFQs raise issues related to the public trust doctrine. In *Illinois Central R.R. Co. v. Illinois* (1892, 453–54) the U.S. Supreme Court found that "the State can no more abdicate its trust over property in which the whole people are interested, like navigable waters and the soils under them, so as to leave them entirely under the use and control of private parties than it can abdicate its police powers in the administration of government and the preservation of the peace."

Because public trust resources are held on behalf of the citizens, the state may be precluded from transferring comprehensive ownership rights to individuals. In general, conveyance of public trust resources to private ownership is a usufruct that does not terminate the state's right to capital or right to manage the resource. Consequently, when a right to harvest fishery resources is conveyed to individuals, the state continues to have responsibility for safeguarding the sustainability of those resources. The Endangered Species Act (ESA), the Fishery Conservation and Management Act of 1976 (FCMA), and international treaties reinforce the state's stewardship responsibilities implicit in the public trust doctrine. McCay (1998) provides an extensive discussion of the application of the public trust doctrine to U.S. fisheries. Macinko (1993) examines the relationship between the public trust doctrine and IFQs. Although the legality of privatizing charter-based recreational fishing has not yet been established, there may be precedence in the regulation and licensing of guided hunting opportunities.

Another option that could be useful in some sport fisheries would be to adopt an annual lottery-based allocation such as that used for many big-game hunting opportunities (Johnston et al., this volume). Because every applicant has an equal probability of receiving a permit and because the number of permits can be set to avoid overexploitation of the stock, a lottery would be unlikely to conflict with interpretations of the public trust doctrine. If lottery winners were permitted to auction their permits, individuals who place the greatest value in sportfishing for a particular species at a specific location would be able to obtain permits. Equity concerns could be partially satisfied by the fact that every applicant would have an equal opportunity of being drawn and that permit sales would be voluntary.

The conceptual appeal of an integrated sport–commercial IFQ management regime is that it would rely on voluntary market transactions to continuously adjust the allocation such that incremental value of quota shares would be equalized between the commercial and charter sectors. In addition, voluntary market transactions would ensure that quota share buyers compensate sellers for the private value of their IFQs. Wilen and Brown (2000) include an excellent discussion of the likely flow of IFQs within and between the commercial and charter-based sport fisheries under various transfer and cap policies considered in the design of the Alaska halibut charter IFQ program (NPFMC 2001). Because the charter-based sport fishery has not been constrained, Wilen and Brown (2000) argue that the incremental net benefit of fish taken in the charter fishery is near zero. Consequently, they predict that if IFQs are allocated to charter operators in proportion to or in excess of their recent catches, charter operators are likely to sell a portion of their initial allocation to commercial harvesters until the marginal private value of IFQs are equated across both fisheries. Rather than focus on the transition to an IFQ-based sport fishery, this example is intended to explore the likely effects of perturbations from that initial equilibrium occasioned by changes in the demand for charter-based sportfishing, moderate changes in halibut abundance, and changes in exvessel demand for commercially caught halibut. Consequently, it will be assumed that the allocation of IFQs between the commercial and charter-based sport fisheries is in initial equilibrium.

In the short run, an increase in the demand for charter-based sportfishing trips will lead to an increase in trip price and an increase in profits to charter operators; total value to anglers beyond the amount they pay for trips will be unchanged in the short run. In the long run, sportfishing charters will purchase additional quota shares, increasing the quantity of trips supplied. Because the purchase of additional IFQs will increase average total costs, charter operators will only purchase additional IFQs if the increase in expected profits is at least as great as the sum of the direct and opportunity costs of acquiring additional IFQs. If sportfishing charters purchase additional IFQs, the benefit to anglers beyond the amount they pay for trips will increase. Because of the added opportunity cost of holding additional IFQs, charter operators will not earn above-normal profits in the long run. Thus in this example, the overall effects of an increase in the demand for charter-based sportfishing trips will be an increase in benefit to anglers beyond the amount paid for trips and a conservation of normal profits for charter operators. In this example, because the sport fishery is initially constrained by their IFQ holdings, there will be no short-run effect on the commercial fishery. However, in the long run, sportfishing charters could purchase additional quota shares from the commercial sector. The key difference between this example and all preceding examples is that, in the present instance, commercial harvesters must agree to the transfer and will only do so if sportfishing charters are willing to pay more for an additional unit of quota than the value of that quota to commercial harvesters. If a commercial harvester sells IFQs into the charter sector, the total commercial quota will decline, causing exvessel price to rise and reducing net benefits to buyers and consumers of commercially harvested fish. If the market for IFQs were initially in equilibrium, the reduction of commercial quota would result in a decrease in profits to the commercial sector. However, for the transfer to be consummated, the charter operators would have had to pay enough to offset the present value of foregone future commercial profits. Decreases in the exvessel demand for commercially harvested fish would have a similar effect, while increases in the exvessel demand for commercially harvested fish would have an opposite effect.

With an integrated market for sport and commercial IFQs, the effects of fluctuations in stock abundance are immediately distributed across both fisheries. Because the sensitivity of the marginal value of IFQs to changes in stock abundance is likely to differ between the two fisheries, abundance changes are likely to lead to short- or long-term transfers of IFQs between the two sectors.

Because the charter operators and commercial harvesters do not benefit from total value to anglers beyond the amount paid for trips or from the net benefits that accrue to buyers and consumers of commercially harvested fish, they will not consider the effect of quota transfers on the magnitude of those net benefits. Consequently, although an integrated sport–commercial IFQ management regime will ensure efficiency from the perspective of commercial harvesters and charter operators, it may not ensure maximization of net benefits to society as a whole.

An Empirical Model of Optimal Sport–Commercial Allocations in the Halibut Fishery

The halibut fishery off Alaska is the only U.S. fishery in the Pacific Northwest region with an IFQ-based commercial sector and a large sport fishery,[3] both of which are economically significant.[4] Halibut are managed under the Halibut Convention of 1923, a bilateral agreement between the United States and Canada. The treaty established the International Fisheries Commission, now the International Pacific Halibut Commission (IPHC). The IPHC is a scientific body with responsibility for conducting stock assessments and recommending conservation measures for halibut in the Pacific Northwest, Gulf of Alaska, Aleutian Islands, and Eastern Bering Sea. In 1976, the Commission's jurisdiction was extended to the 200-mile fisheries conservation zones established pursuant to the FCMA in the United States and similar Canadian legislation. Canadian and U.S. halibut fishers were excluded from each other's territorial and extended-jurisdiction waters in 1978. Although the IPHC retains authority to establish area-specific harvest limits, each nation is responsible within its jurisdictional waters for allocating catches among its various user groups such that the sum of commercial, sport, and other removals does not exceed the IPHC catch limits.

Within the U.S. jurisdiction off Alaska, the North Pacific Fishery Management Council (NPFMC) sets a total allowable catch (TAC) limit for the commercial halibut fishery by subtracting a bycatch allowance and expected sport and subsistence catches from the annual IPHC catch limit.[5] Halibut bycatch mortality has been capped at about 18 percent of the IPHC catch limit in recent years. Subsistence and personal-use halibut catches are small (less than 0.2 percent of the IPHC catch limit). The share of halibut caught by anglers, particularly in Southeast Alaska (IPHC Area 2C) and the Central Gulf of Alaska (IPHC Area 3A: Prince William Sound, Resurrection Bay, Kodiak, Yakutat, Cook Inlet, and adjacent portions of the Gulf of Alaska), has increased from less than 2 percent of total removals in the late 1970s to over 18 percent in the 1990s (Blood 2006). Even before implementation of IFQs, commercial harvesters were concerned that unchecked expansion of the sport fishery would reduce commercial fishing opportunities, particularly in periods of declining halibut biomass. That concern increased substantially after IFQs were implemented. Not only do sportfishing catches reduce the quantity of fish available to individual commercial harvesters in any given year and thus affect commercial profits, expansion of sportfishing affects the wealth of IFQ holders because the asset value of the IFQ is a function of current and expected future catches.

In response to the intensified allocation conflicts between commercial and sport interests, the Council approved establishment of a cap on charter-based sportfishing catches in IPHC management areas 2C and 3A (NPFMC 2000). The cap, called a guideline harvest level (GHL), was set equal to 125 percent of the charter share of the combined commercial and charter catches taken in 1995 (12.35 percent of the combined catch in Area 2C and 15.57 percent in Area 3A), with provisions for a reduction in the GHL if stock biomass declines. Under the GHL, subsistence harvests and harvests by anglers who do not hire guide/charter services continue to be deducted from the commercial TAC. The final rule for the halibut charter fishery GHL was published in the *Federal Register*, vol. 68, no. 153, on August 8, 2003, and the GHL came into effect for the 2004 season. When initially approved by the Council in 2000, the GHL was envisioned as a stopgap measure because it fails to directly limit expansion of the charter fleet and because there was little confidence that traditional sport fishery management measures could effectively constrain catch.

Concerns about the likely inefficacy of GHL management led the Council to approve an IFQ program for the charter sector even before the GHL was implemented (NPFMC 2001). As envisioned, the IFQs that were to have been issued to sportfishing charters would have been transferable between the sportfishing charter and commercial fisheries under conditions intended to provide some stability to both sectors. However, in December 2005, the NPFMC rescinded its approval of the charter IFQ program because of concerns that IFQ allocations based on September 2000 regulations and eligibility criteria would no longer be acceptable politically or legally (NPFMC 2005).

In 2004, the GHL was exceeded by 22 percent in Area 2C and by 1 percent in Area 3A. In 2005, the GHL was exceeded by 36 percent in Area 2C and by 1 percent in Area 3A. Despite some restrictions on harvest by charter crew members, the 2006 GHL was exceeded by 47 percent in Area 2C and by 9 percent in Area 3A. Because implementation of charter effort control measures necessitates amendments to the fishery management plan that cannot be promulgated until at least 2008, it is anticipated that the GHL will again be exceeded in 2007. In June 2007, the NPFMC adopted additional effort controls intended to constrain charter-based catches to no more than the GHL. However, when there is a one-year delay between prosecution of the fishery and generation of data regarding the magnitude of removals and another one- to two-year delay between when the data are available and management measures are selected and implemented, it is unlikely that catches can be constrained to their intended target levels. Because the NPFMC does not expect the GHL will stem a continued de facto reallocation of catches from the commercial fishery to the charter-based sport fishery, an analysis of a moratorium on expansion within the charter fleet and possible development of an IFQ program has been initiated.

Following McConnell and Sutinen (1979), Bishop and Samples (1980), and Edwards (1990), the overall management objective for an integrated commercial and sport fishery can be characterized as a constrained maximization of the net present benefits of commercial fishing and sportfishing over time. The results below characterize the net economic benefits of the set of feasible sustainable commercial sport allocations of halibut in Alaska.[6] Considered independently, the sustainable net benefits of commercial fishing and sportfishing can be represented in figure 6.1.

Profits for commercial harvesters are maximized at a sustainable yield of 42 million pounds and a corresponding biomass of 537 million pounds. The solution that maximizes the net benefits to the combination of commercial harvesters and the buyers and consumers of commercially harvested halibut occurs at a sustainable yield of 54 million pounds and a sustainable biomass of 490 million pounds. That is, when benefits to buyers and consumers of commercially harvested halibut are treated on an equal footing with benefits to harvesters, the optimal harvest level differs from the level that would be optimal from the perspective of a sole owner or a group of IFQ rights holders

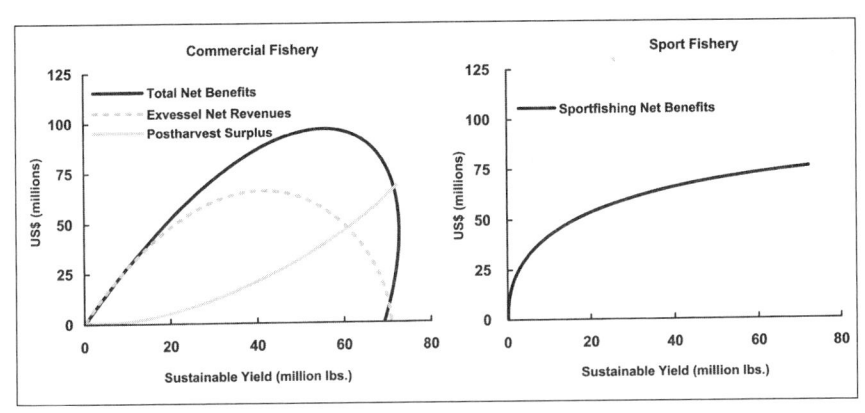

Examining the Interface between Commercial Fishing and Sportfishing 135

FIGURE 6.1
Benefits, Revenue, and Surplus: Commercial Fishery and Sport Fishery
Source: Criddle (2004, 148, 150). Reprinted with permission.

who behave as though they were a sole owner. In a purely recreational fishery, the model suggests that total value to anglers beyond the amount they pay for trips is maximized under a maximum sustainable yield management strategy (72.3 million pounds). But because the incremental net benefits of sportfishing are a declining function of sport catches, the incremental net benefits of catches are quite small for catches beyond about 20 million pounds.

Independent optimization of commercial fishing and sportfishing net benefits fails to ensure welfare maximization because it does not equate the marginal net benefits of commercial fishing and sportfishing. Following Criddle (2004), the feasible set of optimal sustainable commercial–sport allocations and net benefits can be represented by the relationships depicted in figure 6.2.

Because the incremental net benefits of sportfishing exceed the incremental net benefits of commercial fishing at low sustainable yields, most of the sustainable yield is allocated to the sport fishery.[7] As the quantity of fish allocated to the sport fishery increases, the incremental net benefit of additional sport fish catches declines. At intermediate levels of biomass, the incremental net benefits of commercial fishing exceed the incremental net benefits of additional allocations to the sport fishery, and it is optimal to allocate an increasingly large share of the sustainable yield to the commercial fishery. At very high biomass levels, the sustainable yield is small, and it is again optimal to allocate most of the sustainable yield to the sport fishery. The overall optimal solution is to manage for a biomass of 444 million pounds and to allocate 71 percent of the 63 million pounds of sustainable yield to the commercial fishery, with the balance allocated to sportfishing. The overall optimal solution provides US$55.2 million in profit to commercial harvesters, US$26.2

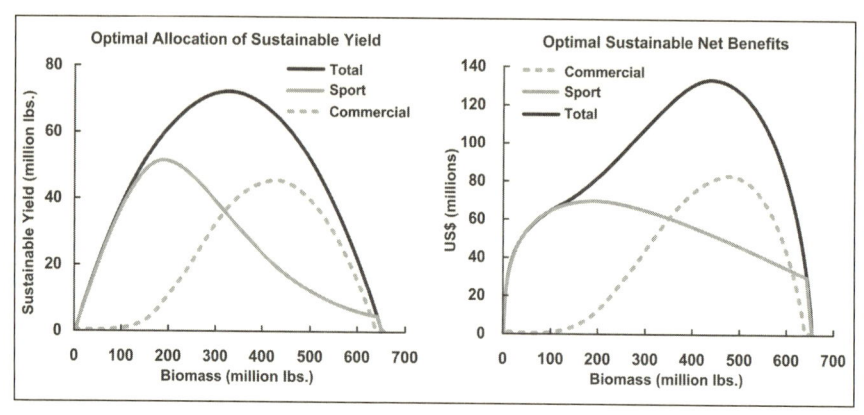

FIGURE 6.2
Optimal Allocation of Sustainable Yield and Sustainable Net Benefits: Commercial Fishery and Sport Fishery
Source: Criddle (2004, 151). Reprinted with permission.

million in net benefits to the buyers and consumers of commercial catches, and US$51.9 million in total value to anglers beyond the amount they paid for sportfishing trips.

Because the social welfare maximizing solution is suboptimal with respect to the private net benefits that accrue to different stakeholders, the actual solution may closely reflect the preferences of the politically empowered. Table 6.1 reflects solutions that maximize commercial profits, net benefits to the buyers and consumers of commercial catches, total net benefits of commercial fishing, total value to anglers beyond the amount they paid for trips, and total net benefits to society by allocating the sustainable yield to the commercial fishery, the sport fishery, or a combination of both fisheries. The maximum sustained yield and open-access solutions are also represented.

The open-access solution represents a fishery managed without consideration of benefits to commercial harvesters, consumers, charter operators, or anglers. Although the open-access solution does not provide profit to commercial harvesters or positive value to anglers beyond the amount paid for trips, it provides substantial net benefits (US$51 million) to the buyers and consumers of commercially harvested fish. Net benefits to buyers and consumers are reduced under management regimes that maximize commercial profits (sole owner or fully efficient IFQ). This result provides an explanation for why commercial fish processors often oppose the implementation of rights-based fishery management programs. In a purely commercial fishery, maximization of the sustained yield generates US$68 million in net benefits to buyers and consumers and results in net operating losses of US$42 million for commercial harvesters.

Table 6.1
Characteristics of Alternative Management Regimes for Commercial Fishing and Sportfishing

Management Regimes	Biomass	Commercial Catch	Sport Catch	Commercial Profits	Buyer & Consumer Benefits	Commercial Net Benefits	Angler Net Benefits	Total Net Benefits
		(million lbs.)			(US$ millions)			
Independent Management								
Open access	444.5	62.7	0.0	0.0	51.1	51.1	0.0	51.1
Maximize sustainable yield	326.1	72.3	0.0	−41.9	67.8	25.9	0.0	25.9
Maximize commercial profits	536.5	42.2	0.0	72.0	23.1	95.1	0.0	95.1
Maximize benefits to buyers & consumers	326.1	72.3	0.0	−41.9	67.8	25.9	0.0	25.9
Maximize commercial net benefits	490.0	54.0	0.0	64.4	37.9	102.2	0.0	102.2
Maximize angler net benefits	326.1	0.0	72.3	0.0	0.0	0.0	75.7	75.7
Joint Management								
Maximize sustainable yield	326.1	37.1	35.2	34.9	17.8	52.8	63.3	116.1
Maximize commercial profits	512.1	37.4	11.4	61.9	18.1	80.0	44.3	124.3
Maximize benefits to buyers & consumers	422.5	45.3	20.6	51.7	26.7	78.4	54.2	132.6
Maximize commercial net benefits	474.5	42.7	14.6	59.5	23.6	83.1	48.5	131.6
Maximize angler net benefits	192.3	8.9	51.2	9.2	1.0	10.2	69.7	80.0
Maximize total net benefits	443.5	44.9	18.0	55.2	26.2	81.4	51.9	133.3

Note: Data based on simulation model developed by the author.
Source: Criddle (2004, 153).

Consequently, it is extremely unlikely that commercial harvesters would voluntarily harvest the maximum sustainable yield (MSY). Total value to anglers beyond the amount they pay for sportfishing trips is maximized when commercial fishing is disallowed and catches approximate MSY. A sole owner of an exclusively commercial fishery would choose to harvest a sustainable yield of 42 million pounds, earning US$72 million in commercial profits and coincidentally providing US$23 million in net benefits to buyers and consumers. If fishery managers were interested in maximizing the total net benefits of commercial fishing, they would set the TAC equal to 54 million pounds and implement regulations to induce commercial harvesters to behave like a sole owner. In so doing, the commercial fishery would generate total net benefits of US$102 million comprised of US$64 million in profits for harvesters and US$38 million in net benefits to buyers and consumers.

The joint management results reported in table 6.1 maximize net benefits to various stakeholder groups conditional on the optimal allocation of sustainable yields between the commercial and sport fisheries. It is notable that the overall optimal solution is suboptimal from the myopic perspective of commercial harvesters, the purchasers of commercial harvests, and anglers. From the perspective of commercial harvesters, the solutions that maximize commercial profits, commercial net benefits, and even net benefits to buyers and consumers are all preferred to the solution that maximizes overall net benefits. Similarly, anglers prefer solutions that maximize total value to anglers beyond the amount they pay for trips, maximize sustainable yields, or maximize net benefits to buyers and consumers. Another important result is that consideration of the joint benefits of commercial fishing and sportfishing provides larger overall net benefits than are generated when the goal of fishery management is solely motivated by an interest in maximizing net benefits to commercial fishers or anglers alone. The results also emphasize the importance of considering net benefits to buyers and consumers.

Because the solutions that maximize overall net benefits, maximize net benefits to buyers and consumers, or maximize the net benefits of commercial fishing produce similar levels of overall net benefits, there are multiple nearly equally efficient solutions with differing distributional attributes. Consequently, even if all of the stakeholders agree to abide by a solution that maximizes net benefits across uses, they will probably contest specific allocation decisions if political–regulatory processes are used to effect the allocation.

Because changes in the demand or supply functions in the commercial fishery, changes in the willingness to pay or cost of participating in the sport fishery, and changes in ocean productivity affect the optimal sustainable yield and the optimal allocation of the optimal sustainable yield, any initially optimal allocation will be suboptimal in subsequent periods. Consequently, to maximize net benefits, allocations need to be modified whenever the economic or biological conditions change. When allocations are determined in a political process, interest groups have an incentive to overstate the marginal value of additional shares. An integrated commercial–sport IFQ management regime shifts the allocation decision from the management arena into the market place. However, the mere act of adopting an integrated commercial–sport IFQ regime will not by itself ensure that overall net benefits are maximized unless the rights are defined in a way that causes the value of net benefits to buyers and consumers and total value to anglers beyond the amount they pay for trips to be expressed in the market price for quota shares. Indeed, failure to account for net benefits to buyers and consumers could be considered to be the basis for the opposition by processors to harvester-only IFQ programs (NRC 1999; Matulich, Mittelhammer, and Reberte 1996; Matulich and Sever 1999). Similarly, to the extent that restrictions on ownership and transfer of IFQs exclude interested parties, efficiency is squandered and political opposition is fomented. The challenge then is to devise a rights-based regime that will encourage expression of the full suite of use and nonuse values. Although it is possible that elimination of ownership restrictions might be sufficient to eliminate the inefficiency of harvester-only rights in the simple case examined in this paper, it is unlikely that such a simple prescription will allow expression of other use and nonuse values.

Conclusion

In commercial fisheries, ownership-by-capture rules lead harvesters to dissipate profits and encourage uneconomic and biologically unsustainable harvest levels. In sport fisheries, the lack of effective limits on the number of anglers and magnitude of sportfishing catches has led to substantial reductions in the commercial TAC and may have reduced overall net benefits to society. Although theoretically possible, the knowledge and control needed to maximize overall net benefit through political–regulatory management regimes is overwhelming and such systems have consistently failed to sustain overall net economic benefits or the resource base on which they depend. When political muscle is the basis for allocating TAC among commercial, sport, and other users, the resultant allocations cannot be expected to maximize overall net economic benefits. The appeal of rights-based management systems lies in their potential to channel rational self-interest in a way that coincidentally maximizes overall net benefit. Because they exploit an alignment of individually rational actions and socially optimal outcomes, rights-based management is potentially self-regulating. If rights can be defined in a way that is meaningful across use and nonuse values and to the extent that all use and nonuse values can be fully captured, self-interest and transferability will encourage the movement of use rights to the use/nonuse that generates the greatest marginal net benefit, ensuring the maximization of overall net benefits.

It is interesting to note that the optimization model suggests that the total net benefit curve is flattened by the inclusion of sportfishing net benefits when compared to the total net benefits solely associated with commercial fishing. That is, when sport and commercial fishing are considered together, there is an increase in the range of biomasses and associated sustainable yields that result in quasi-optimal solutions. It can be anticipated that the recognition of additional use and nonuse values would result in a still broader set of near-optimal combinations of sustainable yield and biomass. Because each of these solutions is associated with different allocations among the claimants and because catch rights without rights that directly affect stock abundance are unlikely to capture some use/nonuse demands and fail to transfer the stock-level choice from the political system, the level of rancor over the choice of biomass and associated sustainable yield can be expected to be intense if the choice is determined through a political process.

With the apparent advantages of rights-based management systems, it seems reasonable to wonder why IFQs, co-ops, and territorial use rights in fisheries (TURFs) have not been warmly embraced and uniformly adopted. The answer is that the creation and enforcement of rights is not costless and that the rights systems that have been proposed often exclude interested parties from ownership; are not structured in a way that is capable of reflecting net benefits to buyers and consumers, total value to anglers beyond the amount they pay for trips, or other use and nonuse values; and do not eliminate the opportunity to free ride on the benefits (costs) of individual stewardship. Because it can be costly to change the legal and social institutions that have been developed to support current fishery management systems and because it can be costly to monitor and enforce quota shares or spatial use rights, especially when there are numerous rights-holders, rights-based management systems will be less prevalent than might otherwise be anticipated (Anderson and Hill 1975; Dennen 1976).

Relaxing the assumptions that there are only two groups of stakeholders and that stakeholders do not value nonuse benefits requires identifying a medium of voluntary exchange that is capable of representing the attributes of the resource valued by the diverse stakeholders. It can be anticipated that when stakeholder groups value fundamentally different aspects of a resource, it may be difficult to define a unit of value that embodies the suite of entitlements and obligations valued by the different stakeholder groups. For example, while capture rights such as IFQs may be a useful medium for voluntary exchanges within and among commercial, sport, and subsistence harvesters, they may be inadequate for reflecting in situ values. Similarly, while fine-scale spatial entitlements might be a more useful medium for voluntary exchange between stakeholders who derive use values and stakeholders who derive nonuse values, spatial entitlements may be an unsuitable medium of exchange for fugitive resources, such as migratory fish stocks. If the resource is involved in important trophic interactions with other resources or if the utilization or conservation of the resource creates external costs or benefits that impinge on the stock or flow of net benefits associated with other resources, finding a meaningful common medium of exchange may be even more difficult. When it is difficult to identify a common medium of exchange to represent the diverse suite of benefits associated with use and nonuse of a collection of interrelated resources, it is unlikely that markets can be relied on to maximize net benefits across the resources and their stakeholders, and it is likely that political processes will remain dominant in the distribution of entitlements among stakeholder groups.

Afterword

The conceptual analysis of the interface between commercial and sport fisheries confirmed that the magnitude and distribution of net benefits from natural resources depends on the rules that determine how access to the resource is allocated between and within groups. While this finding may—and should—seem self-evident, its implications are often overlooked in analyses of so-called optimal allocations of use, nonuse, and option benefits among stakeholders. This finding was intuitive to commercial fishers who received halibut IFQs: as soon as the allocation of catches within the commercial sector came under a market regime, IFQ holders set about influencing fishery managers to adopt measures that would constrain expansion of the sport fishery that threatened to reduce the value of commercial IFQs.

In seeking constraints on the sport fishery, commercial IFQ holders found allies in established sportfishing charters. To these commercial providers of charter services, it was evident that lack of effective constraints on expansion of sportfishing charter capacity, competition with substitute recreational activities, and frequent costly political battles over the allocation of catch shares between stakeholder groups would dissipate profits. Together, these allies suggested the creation of an IFQ within the charter-based sport fishery to constrain the expansion of sportfishing charter capacity and to provide an opportunity to add to or decrease the magnitude of catch shares within the sport fishery in response to changes in the demand for charter-based sportfishing services, changes in the costs of providing charter-based sportfishing services, changes in exvessel prices offered in the commercial fishery, or changes in harvesting costs in the commercial fishery. In addition, supporters of an IFQ management regime for sportfishing charters anticipated that as *de gratis* initial recipients of IFQs, they would be able to earn above-normal profits and that their wealth would be increased by the capital value of the IFQs. While the analysis presented in this paper suggests that sportfishing charters act as perfect competitors, it is likely that they possess some market power and that their support for the implementation of an IFQ-based management regime was, at least in part, due to their anticipation that IFQs would preserve or increase their market power. The recent decision to rescind the charter-based sport fishery IFQ program approved by the NPFMC in 2001 can be attributed to the lengthy delay in development of regulations to implement the IFQ program and to the influence of sportfishing charters who entered the fishery in the interim and would have been excluded from the class of *de gratis* initial recipients under the program approved by the NPFMC in 2001.

Notes

1. The type of sport fishery that will be modeled is one in which anglers hire the services of a vessel and vessel operator for a single eight-hour trip in pursuit of a single species. The vessel operator provides all of the required fishing gear, selects the fishing location, provides recommendations with respect to angling technique, and provides ancillary services related to landing and processing of catch. Providers of these types of sportfishing services are variously described as charter- or headboat operators, or sportfishing guides.

2. The assumption could be relaxed to allow constant levels of bycatch, subsistence catches, and unguided sportfishing catches by specifying that the commercial and charter-based sport fisheries are allocated a fraction of the Optimal Sustainable Yield specified by an external authority.

3. IFQs were adopted in 2007 in one other U.S. commercial fishery where there is also a substantial sport fishery—the Gulf of Mexico red snapper fishery. Although state-managed lake trout and yellow perch resources in Wisconsin's portions of Lake Superior and Green Bay in Lake Michigan are subject to high levels of sportfishing and include IFQ-based commercial fisheries (Muse and Schelle 1989), the commercial sector is very small and may not provide a good example of the interaction that can be expected between an IFQ-based commercial fishery and a competing sport fishery.

4. Gross exvessel revenue from commercial sales of halibut in 2000 was US$157 million (table 16 in Hiatt et al. 2007, 45). Hamel et al. (2002) estimate that more than US$19 million in compensating variation was generated by halibut sportfishing trips in Lower and Central Cook Inlet during 1997.

5. Bycatches are the incidental catches of halibut in non-target fisheries that arise from the imperfect selectivity of the harvest technology. In order to preclude surreptitious targeting, all bycatches of halibut must be discarded unless taken with hook and line gear by individuals with permits to retain halibut. The sablefish fishery is an IFQ-based hook and line fishery. Many fishers hold IFQ for both species and can consequently retain bycatches up to the limit of their quota share.

6. The results are based on an empirically grounded deterministic discrete-time comparative static simulation-optimization of equations (1′) and (2′) derived in Criddle (2004) and appearing in appendix 6.1 at the end of this chapter.

7. It is important to acknowledge that while this result is probably reasonable at high levels of biomass, the data available for estimating total value to anglers beyond the amount they paid for trips did not allow for estimation of the relationship between average trip costs and halibut abundance. Because it is likely that the cost of catching a halibut increases as population size declines, marginal total value to anglers beyond the amount they pay for trips is probably overstated at low abundance levels.

References

Anderson, Terry L., and Peter J. Hill. 1975. The Evolution of Property Rights: A Study of the American West. *Journal of Law and Economics* 18(1): 63–179.

———. 1990. The Race for Property Rights. *Journal of Law and Economics* 33: 177–97.

Bishop Richard C., and Karl C. Samples. 1980. Sport and Commercial Fishing Conflicts: Theoretical Analysis. *Journal of Environmental Economics and Management* 7: 220–33.

Blood, Calvin L. 2006. The 2005 Sport Fishery. In *Report of Assessment and Research Activities 2005*. Seattle, WA: International Pacific Halibut Commission, 49–56.

Cheng, Hsiang-tai, and Oral Capps Jr. 1988. Demand Analysis of Fresh and Frozen Finfish and Shellfish in the United States. *American Journal of Agricultural Economics* 70: 533–42.

Criddle, Keith R. 1994. Economics of Resource Use: A Bioeconomic Analysis of the Pacific Halibut Fishery. In *Proceedings of the Fourth International Symposium of the Conference of Asian and Pan-Pacific University Presidents*, ed. David G. Shaw. Fairbanks, AK: Alaska Sea Grant College Program, University of Alaska Fairbanks, 37–52.

———. 2004. Economic Principles of Sustainable Multi-Use Fisheries Management, with a Case History Economic Model for Pacific Halibut. In *Sustainable Management of North American Fisheries*, ed. E. Eric Knudson, Donald D. MacDonald, and Yvonne K. Muirhead. Bethesda, MD: American Fisheries Society, 143–71.

Criddle, Keith R., and Arthur M. Havenner. 1991. An Encompassing Approach to Modeling Fishery Dynamics. *Natural Resource Modeling* 5: 55–90.

Dennen, R. Taylor. 1976. Cattlemen's Associations and Property Rights in Land in the American West. *Explorations in Economic History* 13: 423–36.

Eales, James, Catherine Durham, and Cathy R. Wessells. 1997. Generalized Models of Japanese Demand for Fish. *American Journal of Agricultural Economics* 79(4): 1153–63.

Edwards, Steven F. 1990. An Economics Guide to Allocation of Fish Stocks between Commercial and Recreational Fisheries. *NOAA Technical Report NMFS 94.* Woods Hole, MA: Northeast Fisheries Center.

Hamel, Charles, Mark Herrmann, S. Todd Lee, Keith R. Criddle, and Hans T. Geier. 2002. Linking Sportfishing Trip Attributes, Participation Decisions, and Regional Economic Impacts in Lower and Central Cook Inlet, Alaska. *Annals of Regional Science* 36(2): 247–64.

Herrmann, Mark, and Keith R. Criddle. 2006. An Econometric Market Model for the Pacific Halibut Fishery. *Marine Resource Economics* 21: 129–58.

Herrmann, Mark, Ron C. Mittelhammer, and Biing-Hwan Lin. 1993. Import Demands for Norwegian Farmed Atlantic Salmon and Wild Pacific Salmon in North America, Japan, and the EEC. *Canadian Journal of Agricultural Economics* 41: 111–24.

Hiatt, Terry, Ronald G. Felthoven, Michael Dalton, Brian Garber-Yonts, Alan Haynie, Kathleen Herrmann, Dan Lew, Jennifer Sepez, Chang Seung, Leila Sievanen, and the staff of Northern Economics. 2007. Real Ex-vessel Value of the Catch in the Domestic Commercial Fisheries off Alaska by Species Group, 1984–2006 (Table). In *Stock Assessment and Fishery Evaluation Report for the Groundfish Fisheries of the Gulf of Alaska and Bering Sea/Aleutian Islands Areas: Economic Status of the Groundfish Fisheries off Alaska 2006.* Seattle, WA: U.S. Department of Commerce, National Oceanic and Atmospheric Administration, National Marine Fisheries Service, Alaska Fisheries Science Center, 45. (Available at http://www.afsc.noaa.gov/refm/docs/2007/economic.pdf.)

Homans, Francis R., and James E. Wilen. 1997. A Model of Regulated Open Access Resource Use. *Journal of Environmental Economics and Management* 32(1): 1–21.

Johnston, Robert J., Daniel S. Holland, Vishwanie Maharaj, and Tammy Warner Campson. 2008. Fish Harvest Tags: An Attenuated Rights-Based Management Approach for Recreational Fisheries in the U.S. Gulf of Mexico. This volume.

Lin, Biing-Hwan, Hugh S. Richards, and Joseph M. Terry. 1988. An Analysis of the Exvessel Demand for Pacific Halibut. *Marine Resource Economics* 4: 305–14.

Macinko, Seth. 1993. Public or Private? United States Commercial Fisheries Management and the Public Trust Doctrine, Reciprocal Challenges. *Natural Resources Journal* 32: 919–55.

Matulich, Scott C., and Murat Sever. 1999. Reconsidering the Initial Allocation of ITQs: The Search for a Pareto Safe Allocation between Fishers and Processors. *Land Economics* 75: 203–19.

Matulich, Scott C., Ron C. Mittelhammer, and Carlos Reberte. 1996. Toward a More Complete Model of Individual Transferable Fishing Quotas: Implications of Incorporating the Processing Sector. *Journal of Environmental Economics and Management* 31: 112–28.

McCay, Bonnie J. 1998. *Oyster Wars and the Public Trust: Property, Law, and Ecology in New Jersey History.* Tucson: University of Arizona Press.

McConnell, Kenneth E., and Jon G. Sutinen. 1979. Bioeconomic Models of Marine Recreational Fishing. *Journal of Environmental Economics and Management* 6: 127–39.

Muse, Ben, and Kurt Schelle. 1989. Individual Fisherman's Quotas: A Preliminary Review of Some Recent Programs. *CFEC Report* 89-1. Juneau: Alaska Commercial Fisheries Entry Commission.

National Research Council (NRC). Committee to Review Individual Fishing Quotas. 1999. *Sharing the Fish: Toward a National Policy on Individual Fishing Quotas.* Washington, DC: National Academies Press.

North Pacific Fishery Management Council (NPFMC). 2000. *Environmental Assessment/Regulatory Impact Review/Initial Regulatory Flexibility Analysis for a Regulatory Amendment to Implement Management Measures under a Guideline Harvest Level and/or Moratorium for Halibut in areas 2C and 3A* (Draft). Anchorage, AK: NPMFC.

———. 2001. *Environmental Assessment/Regulatory Impact Review/Initial Regulatory Flexibility Analysis for a Regulatory Amendment to Incorporate the Charter Sector into the Commercial Individual Fishing Quota Program for Pacific Halibut in Areas 2C and 3A* (Draft). Anchorage, AK: NPFMC.

———. 2005. *Charter Halibut IFQ Motion.* Anchorage, AK: NPFMC.

Roy, Noel, Eugene Tsoa, and William E. Schrank. 1991. What Do Statistical Demand Curves Show? A Monte Carlo Study of the Effects of Single Equation Estimation of Groundfish Demand Functions. In *Econometric Modeling of the World Trade in Groundfish*, ed. William E. Schrank and Noel Roy. Norwell, MA: Kluwer Academic Publishers, 13–46.

Salavanes, Kjell G., and Don J. DeVoretz. 1997. Household Demand for Fish and Meat Products: Separability and Demographic Effects. *Marine Resource Economics* 12: 37–55.

Wilen, James E., and Francis R. Homans. 1998. What Do Regulators Do? Dynamic Behavior of Resource Managers in the North Pacific Halibut Fishery. *Ecological Economics* 24: 289–98.

Wilen, James E., and Gardner M. Brown, 2000. *Implications of Various Transfer and Cap Policies in the Halibut Charter Fishery* (Report). Seattle, WA: U.S. Department of Commerce, National Oceanic and Atmospheric Administration, National Marine Fisheries Service, Alaska Fisheries Science Center.

Case Cited

Illinois Central R.R. Co. v. Illinois, 146 U.S. (1892).

Appendix 6.1

A Theoretical Model of Optimal Sport Commercial Allocations

The overall management objective for an integrated commercial and sport fishery can be characterized as a constrained maximization of the net present benefits of commercial and sport fishing over time. This can be restated as

$$\text{Maximize} NB = \sum_{t=t_0}^{T} \delta^t f(NB_{c,t}, NB_{s,t}) \tag{1}$$

$$\text{subject to } x_t = f(x_{t-1}, h_{c,t-1}, h_{s,t-1}), \tag{2}$$

where $NB_{c,t}$ and $NB_{s,t}$ are the net benefits of commercial and sportfishing, respectively; x_t is the biomass of the target species; $h_{c,t}$ and $h_{s,t}$ are the commercial and sportfishing catches, respectively; and δ is the discount factor. Expanding on a commercial sector model developed in Criddle (1994), a sport fishery model based on Hamel et al. (2002), and an approximate model of population dynamics described in Criddle and Havenner (1991), the net benefits of commercial and sportfishing and the stock dynamics can be described by

$$NB_{c,t} = TR - TC + CS = ah_{c,t} - 1/2 bh_{c,t}^2 - c\alpha h_{c,t}^{\beta_1} x_t^{\beta_2} + bh_{c,t}^2$$
$$= ah_{c,t} + 1/2 bh_{c,t}^2 - c\alpha h_{c,t}^{\beta_1} x_t^{\beta_2} \tag{3}$$

$$NB_{s,t} = \phi \ln(h_{s,t} + 1) \tag{4}$$

$$x_t = \gamma_0 + \gamma_1 x_{t-1} + \gamma_2 x_{t-1}^2 - h_{c,t-1} - h_{s,t-1} \tag{2}$$

where TR, TC, and CS are the total revenues, costs, and postharvest surpluses associated with commercial harvests; and a, b, c, α, β_1, β_2, ϕ, γ_0, γ_1, and γ_2 are estimated parameters. Based on these relationships, and imposing sustainability on the stock dynamics constraint, the model can be rewritten:

Maximize $NB = \sum_{t=t_0}^{T} \delta^t \left(ah_{c,t} + 1/2 bh_{c,t}^2 - c\alpha h_{c,t}^{\beta_1} x_t^{\beta_2} + \phi \ln(h_{s,t} + 1) \right)$ (1')

subject to $h_t = h_{c,t} + h_{s,t} = (\gamma_1 - 1)x_t + \gamma_2 x_t^2$. (2')

The solution to this problem can be obtained from the solution to a related, but unconstrained problem:

Maximize $L = \sum_{t=t_0}^{T} \left[\begin{array}{l} \delta^t \left(ah_{c,t} + 1/2 bh_{c,t}^2 - c\alpha h_{c,t}^{\beta_1} x_t^{\beta_2} + \phi \ln(h_{s,t} + 1) \right) \\ + \lambda_t \left(h_{c,t} + h_{s,t} - (\gamma_1 - 1)x_t + \gamma_2 x_t^2 \right) \end{array} \right]$ (5)

The derivatives of this Lagrangian with respect to the control variables ($h_{c,t}$ and $h_{s,t}$) and Lagrange multipliers (λ_t) provide a set of necessary conditions for an optimum:

$dL/dh_{c,t} = \delta^t \left(\begin{array}{l} a + bh_{c,t} - c\alpha\beta_1 h_{c,t}^{(\beta_1-1)} x_t^{\beta_2} \\ \pm c\alpha h_{c,t}^{\beta_1} \beta_2 x_t^{(\beta_2-1)} / \sqrt{(\gamma_1 - 1)^2 + 4\gamma_2(h_{c,t} + h_{s,t})} \end{array} \right) + \lambda_t \equiv 0$ (6)

$dL/dh_{s,t} = \delta^t \phi(1/(h_{s,t} + 1)) + \lambda_t \equiv 0$ (7)

$dL/d\lambda_t = h_{c,t} + h_{s,t} - (\gamma_1 - 1)x_t + \gamma_2 x_t^2 \equiv 0$ (8)

Whereas the total net benefits of commercial fishing and sportfishing will be maximized when the marginal net benefits of commercial fishing are equated with the marginal net benefits of sportfishing, these nonlinear first-order conditions do not lend themselves to the derivation of an analytic solution. Hence, a simulation-optimization of (1') and (2') is carried out.

Chapter 7

Sport Charter Boat Quota Systems: Predicting Impacts on Anglers and the Industry

James E. Wilen

Proposals to create individual transferable quota (ITQ) schemes in the sport charter fishery have created concerns among sport fishers' groups and charter operators alike (e.g., Bluemink 2005; Loy 2007). Among recreational anglers, there are concerns that individual transferable quotas (ITQs) may cause trip prices to rise unreasonably, that the nature and quality of trips may change in unfavorable ways (e.g., reduced trip flexibility and space), and that abundance of shared stocks available to the U.S. sportfishing sector may be reduced. Among charter boat owners, there are fears that quota programs may reduce the client base, that quota prices may rise and reduce profits, that such programs may involve excessive paperwork and monitoring, and that the recreational sector may lose catch allocations. What should we expect to happen after a quota program is introduced in a sport charter boat industry? Unfortunately, there has been very little analysis of this question.

This chapter speculates on some impacts of concern, using a simple framework that relies on both stylized facts and judicious assumptions that make graphical analysis possible. The next section discusses some stylized facts suggested as starting points in a conceptual analysis of charter boat industries. The third and fourth sections outline open-access and regulated open-access baseline cases that are then used for comparison with the ITQ analysis in the fifth section. The sixth section comments briefly on whether ITQs should be held by owner-operators of charter boats or by anglers, and the final section summarizes and concludes.

Charter Boat Fisheries: Some Stylized Facts

The sport charter industry operates in a competitive market for clients; hence sport charter trip prices are established in a manner that reflects the forces of supply and demand. As a first step in predicting impacts, it is worth decomposing the manner in which supply and demand forces come together to determine current sport charter trip prices. On the demand side, sport charter trips compete with other forms of recreation available to prospective clients. Prospective clients include both residents and nonresidents. Residents and nonresidents may be considered separate markets with different characteristics and hence with different responsiveness to fundamental driving forces. Each is considered in some detail below.

Nonresident Charter Boat Markets

In many charter boat markets, nonresidents are the most important segment of the market. Among nonresident clients, there are typically "walk-ons" who decide to take a day charter trip as a more or less spontaneous part of a trip involving multiple recreational activities and other activities. In addition, many other nonresidents include charter trips as part of a planned package of activities during visits to particular regions for vacation (e.g., Dugan, Faye, and Colt 2007; Herrmann et al. 2001). The total size of the nonresident charter market thus depends upon the population of potential anglers who are visiting a region as vacationers or as clients specifically targeting fishing as an activity.

The total size of the potential market for nonresident sport charter trips is circumscribed by demographic and economic factors. On the one hand, the population of anglers is steadily falling in most regions of the United States (USFWS n.d.), as urbanization eliminates traditions that include rural and subsistence activities from most peoples' experiences. At the same time, as the population ages, more people have leisure time and choose to travel. Nonresident angling is a complement to coastal travel, and travel itself competes with a range of other leisure activities. In this sense, Alaskan or Gulf of Mexico or Florida Keys charter boat trips compete with trips to other places within the country such as hiking trips to Yosemite National Park and wildlife viewing trips to Yellowstone National Park as well as trips to other vacation destinations outside the country such as Canada, Mexico, the Galapagos Islands, and Costa Rica.

The important point is that the potential market for all sportfishing competes within a relatively large number of substitute activities for vacationers' and retirees' dollars and time. What this means is that recreational trip demand in a region is governed importantly by the total market for that region's vacation trips. As a region's vacation trip prices rise, the market demand for trips will fall and vice versa for a price drop. Similarly, as the prices of opportunities to other substitute sites changes, the market for a region's trips will be impacted. If the development of new ecotourism facilities in Costa Rica increases and trip prices there fall, we would expect that at least some potential visitors to Florida or the Gulf of Mexico to divert some of their vacation budgets to Costa Rican trips. At the same time, as habitat losses in Pacific Coast state salmon streams reduce opportunities for citizens of Washington, Oregon, and California to fish in their own backyards, some budgets will make room for substitute trips to Florida and the Gulf.

In addition to the fact that each region's trips compete in a North American and world market for vacation experiences, there are other fundamental driving forces affecting market demand over the long run. One of the most important is the health of the economy, reflected particularly in household income levels. When times are good, the market for leisure and for recreational and vacation trips will be strong; when economic activity wanes, market demand will fall.

While demographic and economic factors condition the size of the larger market for leisure activities, local recreational charter boat markets depend upon prices for the charter boat experience and prices of other substitute experiences. For nonresidents, however, a change in charter trip prices is likely to represent only a small percentage change in the total trip cost. This implies that we can invoke Marshall's "importance of being unimportant" to conclude that the charter demand for these individuals is relatively insensitive to trip price. In other words, adding US$20 to a US$125 single day charter trip price taken by a nonresident who has spent several thousand dollars on a vacation trip to a region will cause some reduction in the demand for charter trips at the margin, but the impact is not likely to be large. Hence a stylized fact that will be maintained in the following analysis is that the charter trip demand by nonresidents is relatively price insensitive.

Resident Charter Boat Markets

The demand for resident charter boat sportfishing trips is driven by the same fundamental forces, namely the price of trips, the money and time price of substitute activities, incomes, and preferences for fishing and other outdoor activities. A factor that distinguishes much of the resident demand from nonresident trip demand is the role played by the opportunity for fishing to fulfill what we might loosely call subsistence needs. Many resident fishers live close to the coast precisely because of the opportunity to be partially self-sufficient in the provision of food needs with fish. For these individuals, a sport charter trip represents the opportunity to not only have a desirable recreation experience but also to fill a freezer with fish. But subsistence-oriented individuals typically engage in other recreational opportunities to fill their freezer with substitute foodstuffs, including game, other fish, crab, etc. Hence other local opportunities to engage in subsistence activities in a particular region compete with salt water charter opportunities. As a result, we expect, as another stylized fact, that the demand for trips from residents is relatively more price sensitive than it is for nonresidents.

In addition, trip demand for resident charter boat clients is also likely to be more responsive to the chance of success. Thus, when particular charter boat targets of fish are abundant and fishing is abnormally good, residents will be more likely to be responsive for two reasons. First, residents are likely to be plugged into the network of informal local information sources about relative catch rates. Second, residents are more likely to be able to respond to high success rates by juggling time availability and conflicts with other uses of leisure and work times. Our third stylized fact for our analysis below is that resident demand is more responsive to catch rates and other measures of success, at least in the short run.

The Industry under Open Access

How would we expect the market for charter trips to equilibrate under an open-access system? To answer, we must make some assumptions about the size of the sport fishery relative to the growth potential of the resource. In addition, we need to make some assumptions about how regulations are utilized in the sport fishery. Suppose, to begin with, that recreational fishing mortality is insignificant relative to natural and commercial mortality and that regulators simply account for recreational harvest by subtracting it from the total allowable catch (TAC). In this case, nothing limits total harvest, and there is no feedback between recreational mortality and abundance or regulations. In panel A of figure 7.1, a market is depicted that serves both resident and nonresident anglers. In this figure, stylized facts are incorporated by depicting the demand from nonresidents as relatively price insensitive whereas the demand from residents is price sensitive. Note that the nonresident demand line labeled D_{NR} has a much steeper slope than resident demand line labeled D_R, indicating the relative insensitivity to price changes by nonresident anglers. The combined demand from both markets is depicted with the heavy line, which sums the two demand curves horizontally at every price.

Also depicted is the marginal (and average) cost curve labeled MC* associated with operating the charter fleet. MC* (assumed equal to the average cost

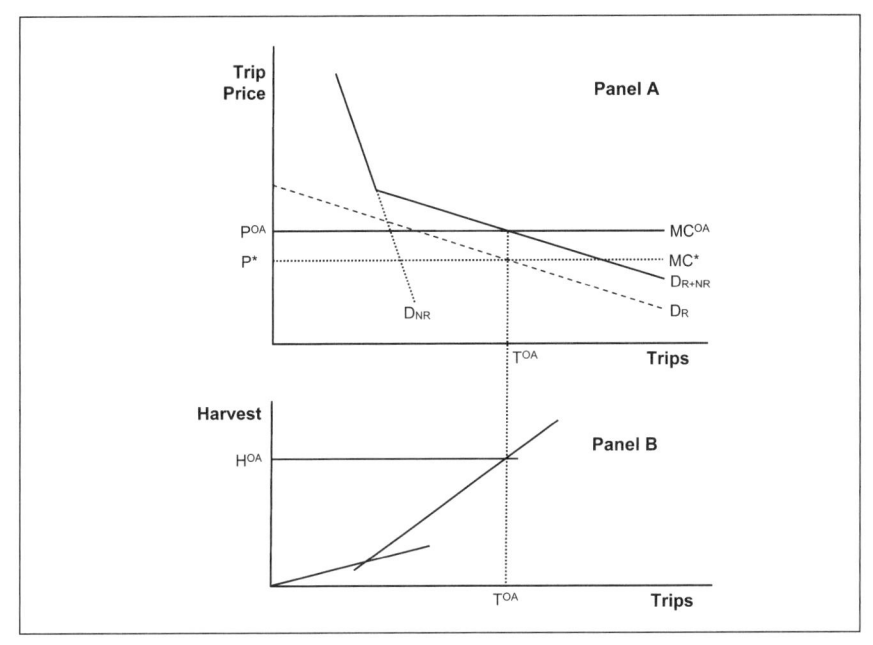

FIGURE 7.1
Open Access

curve AC*) represents costs per trip that would be incurred with an efficiently operating industry with secure property rights. These are the lowest costs associated with providing recreation trips, and they presume reasonably fully utilized capacity and efficient use of variable inputs. An efficiently operating industry would drive trip prices down to P*. In contrast, MC^{OA} represents the open-access marginal and average cost curves, assumed to be higher than MC* to reflect the relative input inefficiency of open-access production conditions. These higher costs might incorporate the additional costs associated with having too many vessels, each operating at excess capacity. They might also incorporate wasteful variable input use, such as competition between firms in a race-to-fish.[1] The open-access industry equilibrium occurs at a trip price P^{OA} and total trip supply of T^{OA}. Trip prices are higher than P* under open-access conditions because costs are higher and the industry attracts fewer anglers than would be the case under efficient conditions with secure rights to the resource. Importantly, residents and nonresidents take fewer trips under open access at higher prices than would be the case in an efficient industry. Because the equilibrium price P^{OA} is higher under open access than the efficient price P*, total angler consumer surplus[2] from both groups is lower than it would be under efficient conditions.

Under the circumstance in which the harvest taken by the charter industry is unconstrained by regulators, this open-access equilibrium in the charter industry will result in a certain total harvest H^{OA}, shown in panel B of figure 7.1. What is this amount? It is simplest to assume that total harvest will be proportional to trips taken by individuals in each market, holding abundance constant. This is borne out in practice. It is also reasonable to assume that nonresidents are generally less skilled than locals, ceteris paribus. This is another stylized fact that is borne out in practice: most nonresident anglers take home fewer fish per trip than residents.

We depict total harvest on the cumulative harvest graph in panel B of figure 7.1. The first segment represents cumulative take by nonresidents (at a constant rate per trip) and the second segment represents the incremental take by residents (at a higher rate per trip indicated by the steeper slope). This market production function is drawn under fixed abundance conditions that determine landings and harvest per unit effort, assumed to be proportional to total trips by each group. It also is drawn under some implicit assumptions about the manner in which trip demand is affected by catch rates. Trip demand curves and corresponding harvest functions are assumed to be equilibrium relationships, depicting a total harvest of H^{OA} associated with a total trip demand of T^{OA}.

Now, what happens when abundance rises in this system? Basically the catch rate per trip will rise, twisting up the harvest functions that are assumed proportional to trips. Then, to the extent that rising catch rates shift trip demand, the new equilibrium will involve higher demand, ceteris paribus. Since resident demand is most responsive to catch per trip, we would expect higher abundance to generate a shift out in resident demand for trips, D_R, reflecting the added satisfaction generated by higher expected catch per trip. We would expect much less response to higher abundance from nonresidents. Under these assumptions, the new equilibrium would occur at the same price per trip (because we are assuming that prices are determined by the constant marginal costs MC^{OA}) but with more trips taken. Total harvest would rise, and the increment would be taken mostly by increases in the numbers of catch per unit effort (CPUE)-responsive resident angler trips. The overall result of increased abundance under open access would thus be higher demand for trips, more trips taken, more angler consumer surplus, and more harvests.

These results represent the open-access scenario in which total harvests in the recreational sector endogenously increase or decrease according to the general factors that influence the market for charter trips by residents and nonresidents, as well as to the role played by fish abundance. In this scenario, there is no constraint on total harvests in the charter sector since it is assumed that either there is no commercial sector or that sport charter allocations are subtracted from the overall TAC before commercial harvest limits are determined. As economic conditions change, be they incomes, prices and availability of substitute trips, or demography, the demand curves may shift with the size of the overall market for sport charter trips. As abundance changes, we also would expect some changes in participation by those responsive to expected harvest, which in turn is related to abundance.

Total recreational sector harvest is also governed to some extent by industry convention with regards to catch and release fish. A recreational trip by a walk-on nonresident may involve an average of one fish per trip, even when the bag limit is two, because such tourists do not particularly want to keep two fish. On the other hand, some residents may wish to maximize the pounds kept per trip, up to limits determined by the bag limit. Hence their harvest rates per trip may reflect strategies involving decisions about which fish to keep as one's first fish, whether to throw back subsequent fish below certain sizes, pooling fish over several individual limits, etc. Importantly, however, in this open-access scenario the total harvest is determined as a consequence of the market for trips rather than vice versa. This story changes if recreational regulatory constraints become binding for the industry as a result of either more restrictive bag limits or season limits under regulated open access (or a quota system).[3]

The Industry under Regulated Open Access

There are two main issues that are important in thinking about the impacts of typical regulated open-access constraints. The first is the likelihood that the measures that are designed to be triggered will, in fact, become binding on the industry. The second issue is whether the promise of binding measures can induce changes in charter boat operator behavior in anticipation of those constraints. In some regulated charter boat fisheries, in contrast to the open-access story above, an overall TAC for the sport sector is determined by regulators. Then, some policy instruments must be used to ensure that the targeted TAC is actually not exceeded. Measures that are believed to be most effective in actually constraining harvest are more restrictive individual (or vessel) bag limits per trip. Season length restrictions might also constrain aggregate harvest, although we would expect some compensating behavior that would involve intensified activity during the remaining compressed open season. Shorter seasons would probably disproportionately affect operations catering to tourists, since the tourist market is less capable of adjusting around season length restrictions.

Consider first the implementation of more restrictive bag limits. In principle, it would be possible to reduce total charter sector harvest by reducing bag limits, say from two to one fish per person per trip.[4] In practice, however, it is difficult to actually fine tune bag limits to the degree that is needed to squeeze aggregate harvest. The first problem is with the discreteness of bag limits (i.e., the fact that they must be implemented in integer units such as one or two fish per person per day). The discreteness of bag limits reduces their flexibility and ability to smoothly reduce harvest. For example, in the Alaskan Kenai sport charter boat halibut fishery, catch typically averages about 1.8 fish per angler per day, with about two additional fish caught and released (Herrmann et al. 2001). Adopting a one-fish bag limit would likely affect nonresident and resident anglers differentially, depending upon how much of the value of the experience was related to catch and release or to putting fish in the freezer. Predicting how much reduction in harvest will take place as result of a bag-limit reduction is thus an empirical question related to how much the demand shifts and how that shift affects industry demand and supply in equilibrium.

A second issue is related to the fact that a one-fish bag limit reduces the charter boat ability to spread a bag limit across all customers. It is rarely the case that each client hooks and lands a fish; hence it is always possible to share fish across all clients without exceeding the vessel bag limit. With a two-fish limit per person there is considerable flexibility to do this; with a one-fish bag limit, it would be much more difficult, probably reducing trip demand by both residents and nonresidents.

An issue with regulated open-access policies is whether the anticipation of the triggering of either bag-limit or season length reduction policies can induce behavior in such a manner that they are not actually triggered. In many ways, it would be ideal if charter operators could induce clients to voluntarily reduce their retention of fish, without necessarily reducing the actual catch rate. Thus in periods of pending binding constraints on total industry take, emphasis might be shifted to the sport of hooking, landing, and releasing fish rather than harvesting them. This could be done by charter operators announcing that trips would henceforth be one fish per person trips (for the remainder of the season in which the aggregate harvest constraint threatens to bind). If charter operators could successfully achieve voluntary one fish per person per trip behavior, then the bag or season limits triggers might be avoided. We would expect some change in strategic behavior among fishers as a result, however. With a one-fish limit, there would be more of an incentive to release small fish caught initially in hopes of landing larger fish. This might offset the intent of the harvest reduction somewhat, both by raising the average size of landed fish and by increasing hook mortality.

In the final analysis, the problem with relying on voluntary changes in behavior is that there must be some mechanism to enforce behavior among participants who collectively stand to gain from cutbacks, but who individually

Sport Charter Boat Quota Systems

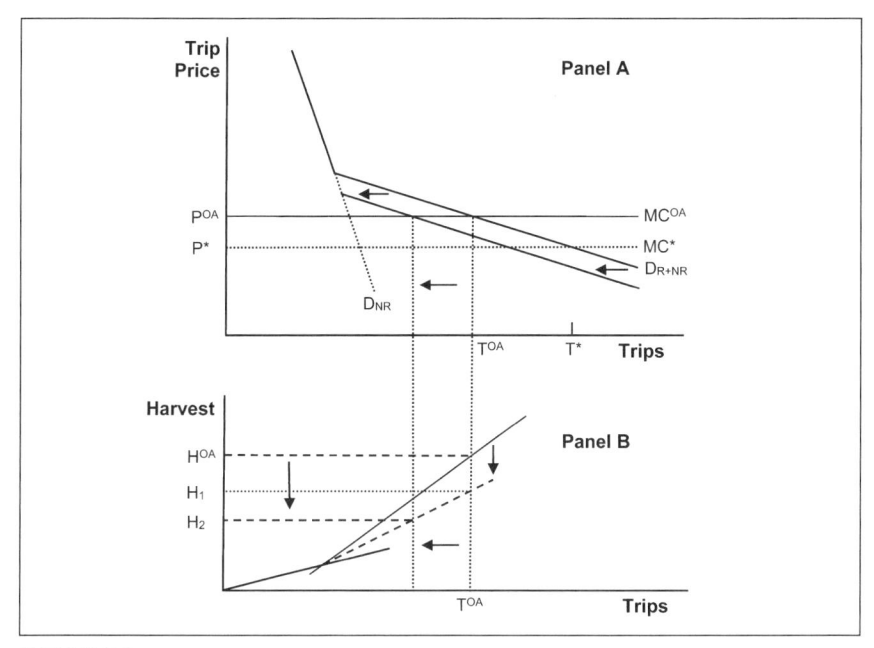

FIGURE 7.2
Recreational TAC

stand to gain from deviating from the collective policy. We might anticipate that charter operators would self-enforce, by watching each other and disenfranchising those who fail to follow the voluntary one-fish policy.[5] Unfortunately, it is equally likely that charter operators would either choose to look the other way or to actually hide fish caught in excess of the informal one-fish rule. This latter possibility is something that managers should truly worry about, since the data base would then be distorted in unknown ways. It may seem a paradox, but it is certainly possible that attempts to induce voluntary cutbacks in landings might actually only produce a reduction in reported (but not actual) landings.

Figure 7.2 with panels A and B shows some possible impacts of either regulated or effective voluntary bag-limit reductions. We assume first that there is a need to cut back total harvest from H^{OA} to H_2 via bag-limit reductions. We depict this, for simplicity, as affecting the resident market only, which we assume to be more bag-limit responsive. First, there would be some impact on the harvest rate per resident trip, shown by the twist in the harvest rate function downward, shown in panel B. This shows that each resident trip yields a smaller harvest rate after regulated or voluntary reductions in harvest are implemented. If trip demand remained at T_{0A}, the aggregate harvest would fall to H_1, which is a fraction of the desired reduction. But with lower bag limits, the demand for resident trips would actually fall somewhat, shown by the leftward shift in the combined (heavy dark) market demand curve D_{R+NR}. Some combined equilibrium reduction in take per trip and reduced resident demand would achieve the desired reduction in aggregate harvest to H_2, at a cost of reducing both the quality and quantity of resident angler trips.

In summary, under regulated open access, the industry impacts of regulations depend upon how specific policy options affect both the quality of the recreational experience and fishing harvest. More restrictive bag limits or seasonal closures have the direct effect of reducing fishing mortality but also an indirect effect associated with reducing the demand for trips. In addition, bag limits are likely to alter the choices that fishers make regarding how many fish are released, the size and sequence of releases, and, as a consequence, the discard mortality. While the impacts on discard choices are beyond the scope of this chapter, they are likely to be very important in fisheries where discard survival is low.[6]

The Industry with ITQs

What impact would a transferable quota program have on a charter industry and how would it affect anglers in the charter market? We will assume that quotas are held by charter boat operators and not by anglers themselves. In this case, the most important effect of ITQs would be that they would change fundamental incentives faced by charter operators. In particular, with a guaranteed quota, charter boat operators would shift some of their attention toward trying to maximize the values derived per pound of quota held. Exactly how that would unfold is not clear, since it depends upon the nature of the inefficiencies generated under open access.[7] This issue is unresolved in the literature on conventional commercial fishing, and it has been given much less attention, if any, for recreational fishing. In what follows, the various kinds of specific open-access and regulated open-access inefficiencies are assumed to be encompassed in the open-access cost function for trips.

Panels A and B in figure 7.3 show the "first round" impact of ITQs on the charter boat system, assumed to be before trade and consolidation occurs. In this figure, we assume that the charter industry is granted a supply of quota H^{OA} equal to the amount that was previously caught under the open-access status quo setting. This is shown by the horizontal total harvest curve in panel B at an aggregate ITQ level equal to H^{OA}. We assume that, in the first round, each charter boat operator is granted allocations equal to his/her pre-ITQ harvest levels, and that these are not initially traded.[8] Under these circumstances the initial impact of an ITQ system will be to induce charter operators

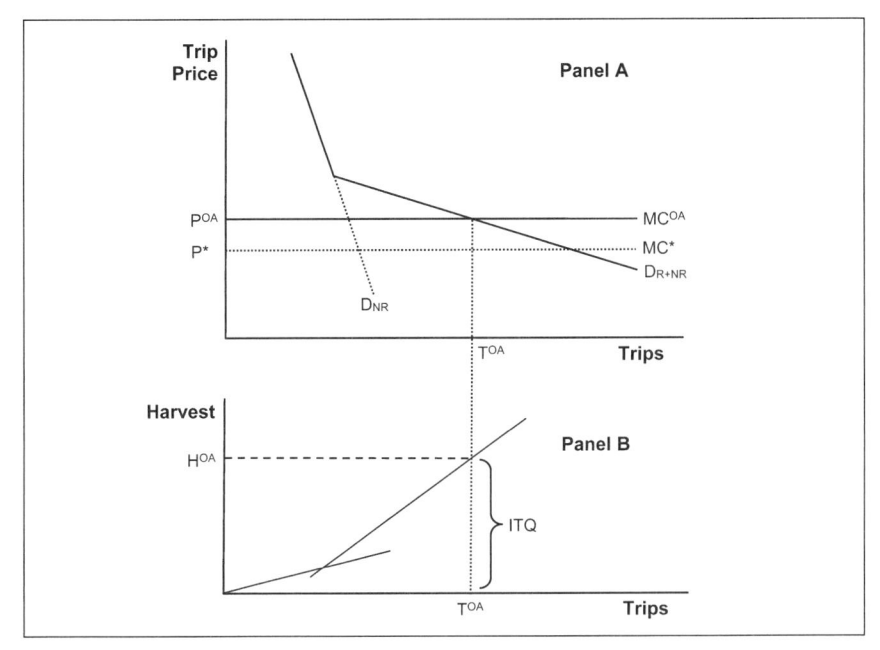

FIGURE 7.3
ITQs—Initial Impacts

to rearrange inputs, management, and fishing practices in order to reduce the costs of trips. This is reflected in the marginal costs of operation falling to the efficient level MC* from MC_{OA} in panel A. The kinds of changes likely to be induced by ITQs are many, including less time spent searching, more trips scheduled at full capacity, and rationalization over the season so that more trips are taken during high yield periods.

Initially, then, operators in the industry will make changes in operations that save the additional costs that were incurred under open-access conditions by an industry competing for the fishery resource without secure rights to it. ITQ quota prices will take on positive values as costs are reduced, reflecting precisely the additional profits generated as a result of the cost saving. Quota prices will rise to a level such that the total lease value in quotas is roughly $[(P^{OA} - MC^*)*T^{OA}]$ and the price per pound is $[(P^{OA} - MC^*)*T^{OA}]/ITQ_{pounds}$. Note that the price of trips does not rise to "pay" for ITQ prices. Instead, at least initially, the market clears at the open-access price and quota prices are driven up to the gap between the market-driven price of trips and the new lower cost of trips.

First round impacts are assumed to occur before ITQs are traded and are associated with the initial cost reductions generated under the new incentive regime provided by ITQs. As it turns out, cost reductions are only part of the changes expected after introducing ITQs. The other category of changes would be those associated with changes in the market side of resource rents. In particular, holding quotas not only generates incentives to reduce cost per unit quota used but also incentives to increase revenues per unit quota held (Homans and Wilen 2005). An important feature associated with the fact that quotas attain value is thus the incentives that quota values in turn create to increase those values.

What would we expect might happen in the marketing of sport charters under assumed stylized facts as a result of quotas? Some insight into this question can be gained by noting first that a charter boat quota holder has an incentive to court clients who are actually inefficient in catching fish. If all clients pay the same price for a trip, it pays the charter boat operator to try to induce clients who won't use much quota per trip to take a trip on his or her vessel.

While the proportions of resident and nonresident clients differ in different ports with recreational charter industries off coasts, in most ports the market serves some of both types of clients. If it were possible to charge residents and nonresidents different prices, one might see some competition among charter boat owners for the lower efficiency, nonresident anglers. One can conceive of situations, for example, whereby charter owners advertise in tourist magazines or arrange vacation packages tied with lower trip price subsidies to visitors to popular winter destinations like the Gulf of Mexico, etc. In the long run, however, it is actually difficult to discriminate between the different types of clients by charging different trip prices. However, it is possible to segment the market into different groups differing only by harvesting efficiency. The easiest way to do this is to charge different prices to each person according to his/her own harvesting efficiency. And the best way to do that, of course, is to charge a fixed price per trip, with an additional price per fish taken home. Hence it seems likely that the ultimate way for the industry to maximize its revenues associated with quota held is to charge clients for fish taken, in addition to a trip fee covering operating costs. An important conclusion is thus that it is likely that the pricing system in a recreational fishery with ITQs will transform into a two-part system with a fixed price per trip and an additional charge based on harvest that covers the cost of providing quota.

Unfortunately, it is more difficult to depict in simple graphs the manner in which the industry might ultimately equilibrate with a combination of trip price and harvest quota price system. With a flexible pricing system in which trip prices and harvest prices can be set separately, charter clients essentially face two prices. Under this system each individual will choose both the quantities of charter trips taken over the year and the quantity of fish kept simultaneously in a manner dependent upon relative prices. The market demand for trips is thus dependent on the price of quota (P_Q) for harvested fish in addition to trip prices (P_T) and other factors such as income (Y). Similarly, the demand for quota to cover harvested fish is dependent upon the price of trips in addition to the price of quota and other factors. We can write these market demands as $T^* = T(P_T, P_Q, Y)$ and $Q^* = Q(P_Q, P_T, Y)$. The characteristic of the market with a two-part pricing system that distinguishes it from either the open-access or a bag-limit-enforced overall guideline harvest level (GHL) system[9] for a recreational sector is that in the quota system, the amount of fish actually harvested per person per trip will not be simply a given quantity related to relative abundance through catch per unit effort (CPUE). Instead, the number of fish landed will be a choice made by each angler according to his or her preferences for landed fish, given the price of quota that must be paid to cover the quota lease price. This de-couples the two panels of our graphs in figures 7.1–7.3.

In figure 7.4 an equilibrium in the ITQ market is depicted after implementing two-part pricing by showing two pairs of market demand curves and the corresponding harvest production functions for each market group

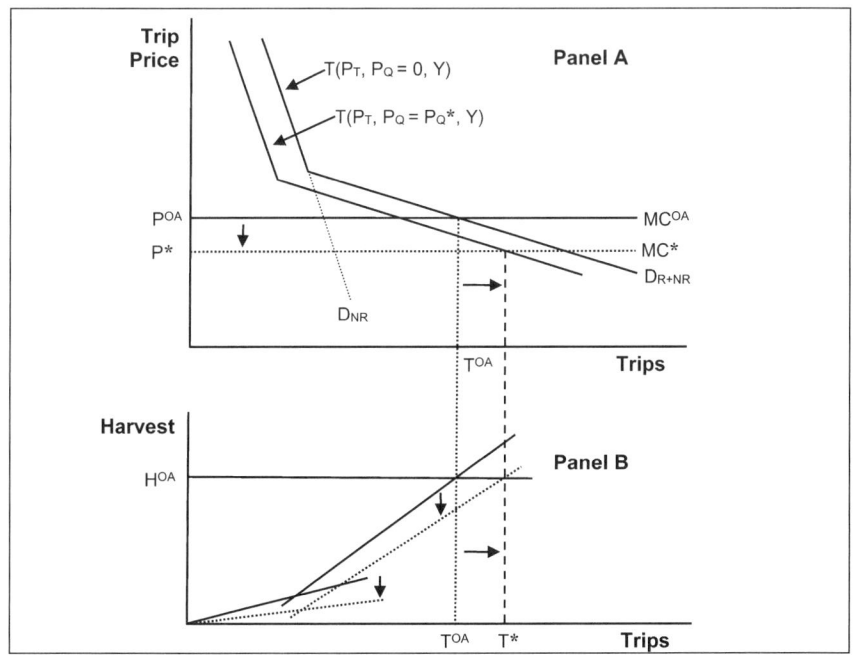

FIGURE 7.4
ITQs—With Trade

in the lower panel. In previous graphs, the panel B harvest production functions were depicted as involuntary and only a function of relative abundance (CPUE) and skills in each market group. In the ITQ case, the slopes of the two segments of the production functions depend upon choices made by resident and nonresident anglers in the face of having to pay prices per pound for each landed fish, and thus are endogenous.

Figure 7.4 is drawn with the right solid line labeled $T(P_T, P_Q = 0, Y)$ reproducing the first round equilibrium in figure 7.3, which depicts the market before the two-part pricing scheme is adopted (hence $P_Q = 0$). The left solid line in figure 7.4 labeled $T(P_T, P_Q = P_Q^*, Y)$ shows the eventual equilibrium after two-part pricing is implemented. In panel A, the market demand curve is shifted leftward because trip demand depends upon the price of quota. The right solid line represents market trip demand curves under the assumption of a zero quota price (the conditions depicted in figure 7.3); whereas the leftward shifted demand curve represents the ultimate trip demand curve with non-zero quota prices. In panel B of figure 7.4, I depict the voluntary equilibrium harvest rate choices made by clients who must now pay for each fish caught. The dotted line harvest rates are drawn below the involuntary solid line curves under the assumption that both groups will choose to harvest fewer fish per trip once they have to pay for the fish.

In the full long-term equilibrium, the charter industry will equilibrate in a manner in which the trip price is driven down to the level covering operating costs, or $P^* = MC^*$. As shown in figure 7.4, whether total trip demand increases in the final equilibrium depends upon the interplay between trip prices and quota prices. Higher quota prices will reduce the total market for trips (shift the demand curve leftward) but a lower trip price will compensate by inducing more trips. In panel B, voluntary choices of lower harvest rates are associated with non-zero costs of landing a fish, and this shifts the harvest functions down. As we have drawn the graphs in the two panels, total trips increase because trip demand by residents is reasonably responsive to trip price, but trip demand does not fall dramatically as quota prices attain non-zero levels. Harvest rates per trip do fall also as a result of anglers being charged positive quota prices to land fish, and this allows more trips T^* and anglers to participate in the industry for the given allowable harvest H^{OA}.

The above analysis assumes that quotas are transferable, if not between the commercial and charter sectors, then at least within the charter sector. Without the ability to consolidate excess vessels to remove surplus capacity, some of the cost cutting that would be passed off to clients will not take place. This mismatch of initial and final efficient capacity is likely to be particularly significant if the lead-in period between announcement and implementation allows entry of individuals hoping to gain a share of the quota. Without transferability, there would still be incentives to operate in cost-efficient manners and to add value to quotas held via the market for charter services. In addition, other experience has shown that entrepreneurs devise alternative methods to transfer, lease, or otherwise sell quota, even when de facto sales are not allowed.

Allocating ITQs to Anglers or Guides?

Some have raised the issue of whether quota ought to be held by anglers (clients) rather than by charter boat operators. One can envision an angler-held ITQ scheme in theory, but there are implementation and operating difficulties that would have to be overcome in practice. In fact, other sport activities operate in this manner, including many big game hunting activities that involve limited numbers of harvest permits (e.g., see Johnston et al., this volume; Leal and Grewell 1999). In most of these, hunters draw permits to harvest in a lottery, and guides then take the hunters out to attempt to meet their bag objective. The guide market is competitive, and each guide offers services that include tangibles related to the comfort of the experience (food, tents, clothing and equipment, transportation) as well as services relating to the harvest success. Something like this kind of scheme could be used in a charter fishery, and the lessons learned in operating such a scheme could be drawn from big game hunting. One design issue is related to whether a permit should be for a fish or for pounds of fish. If permits were per fish, we would see "high grading," with discard mortalities going up, and a monitoring scheme would be needed to account for discard losses in the computation of the charter sector share. If permits were for pounds of fish, there would need to be some secondary market or other scheme for sweeping up odd lots of unused quota and for attaining excess quota to clear the market.

An advantage to leaving quota in the hands of charter operators is that they are more likely to be cohesive and effective spokespersons for resource stewardship. Since their livelihoods are tied directly to habitat and stock conditions, it is in their interest to both monitor biomass health and speak out if conditions deteriorate. While it might be possible for representatives of angler quota holders to speak on their behalf, quota ownership would vary from year to year, reducing the continuity that comes from permanent ownership. One consistent lesson that arises out of experience in other ITQ programs in commercial fisheries is that property rights generate significant changes in attitudes toward long-term stewardship (e.g., see Griffith 2007; Wilen 2006; Sharp 2005). To the extent that program design can encourage these, they make the management task ultimately easier.

Conclusion

A significant issue of concern among representatives of recreational anglers or clients of charter boat operators is exactly how the experience and satisfaction of the charter trip will change after ITQs are implemented. This is understandably a concern since ITQs will induce changes in the structure of the industry that are difficult to predict at this juncture. An important point is that the comparison needs to be made against the alternative system that would prevail if ITQs were not implemented. The graphs used to discuss the alternatives give some insight into these questions. If we begin with an unregulated open-access fishery, an important characteristic of that system is that there is over-capacity inefficiencies associated with the open-access nature of the industry. In commercial fisheries this manifests itself in too many boats, with too much capacity, chasing too few fish. In a sport charter fishery it is likely that there are also too many boats, taking trips at less than full capacity, using perhaps too much effort finding and landing fish. The consequence of this for the client anglers is that the trip price must cover these inflated costs and hence trip prices are relatively high (P^{OA} in figure 7.1). A rationalized fishery in which these inefficiencies were removed would actually serve more angler trips and at lower trip prices. Unregulated fisheries thus generate a utility and welfare loss associated with open access that is paid by anglers as a group, in the form of higher trip prices and fewer trips taken.

If the charter sector total harvest is constrained by either bag-limit-enforced general harvest guidelines or season length restrictions or by an ITQ program, the constraints on harvest will also have some utility cost to anglers relative to the circumstance in which the charter sector is allowed to expand catch at will. This can also be seen as changes in consumer surplus under the figures showing different program impacts. For example, consider what happens under an increase in abundance that shifts the trip demand function out. Several scenarios are possible. Consider first a policy that holds total harvest at H^{OA} in the face of the abundance increase instead of granting a proportional TAC increase to the charter sector. If this is done with restricted seasons that keep total trips constant, then trip prices will have to rise above P^{OA} so that the increase in demand does not increase total trips and harvest mortality. Then there is a loss in consumer surplus associated with the fact that trip prices are driven up, offset somewhat by an increase in consumer surplus associated with the higher satisfaction per trip due to higher CPUE.

In contrast to a binding recreational TAC, figure 7.2 shows what happens under an alternative policy that induces anglers to voluntarily restrict harvests. This might be done by allowing clients to keep fish only above a certain size, or to keep only a specified total poundage of fish. The important point illustrated in figure 7.2 is that these "voluntary" reductions actually reduce the utility derived per trip; hence there is a consumer surplus loss associated with the leftward shift of the demand curve. With these kinds of policies, the restriction can be achieved without altering trip prices, but there is clearly a reduction in angler satisfaction and a resulting reduction in the demand for trips relative to the situation without the regulated open-access controls.

With an ITQ system in place, there are two impacts on angler client welfare or satisfaction. First, there is an increase in consumer surplus associated with the reduction in trip prices to levels that cover the reduced operating costs after more efficient practices are adopted. If nothing else were changed, this price reduction would expand the charter market, and there would be more angler trips taken. But there is an additional impact if the system evolves into one that charges harvesting fees to clients based on the ITQ opportunity costs. These additional harvest charges can be viewed as increases in the price of complementary inputs to the recreational experience, and they will reduce the demand for trips as a consequence. Overall, then, an ITQ system will increase client welfare by reducing trip prices, but this will be offset by welfare reductions associated with paying non-zero prices to harvest fish. How these impacts net out is likely to depend upon empirical relationships that reveal more completely how trip demand and willingness to pay are affected by catch rates, harvest rates, and other characteristics of the experience.

Notes

1. As is the case in a commercial fishery, a race-to-fish can occur in a recreational fishery when fishing closures are used to control total recreational fishing mortality.

2. Consumer surplus is the total value to anglers of a fishing charter beyond the price they pay for a trip. It is the area above the cost curve and below the joint demand curve up to the intercept.

3. The concept of regulated open access depicts a situation in which entry is open, but all participants must adhere to some regulations. Generally, regulations aim to ensure that biological health is maintained and take forms that limit fishing mortality, such as season closures. See Homans and Wilen (1997).

4. This was actually proposed for some regulatory areas in the Alaska halibut recreational fishery in 2007 by the International Pacific Halibut Commission, but it was not carried out by U.S. managers. See Baeth (2007).

5. Disenfranchising can occur in various ways. For example, charter operators often share information on catches and locations during the day. Operators may refuse to share such information with operators who have been disenfranchised.

6. These are issues we address in Abbott, Marahaj, and Wilen (2008), and Abbott and Wilen (2008).

7. For examples of the different ways quota value are increased see Wilen (2005 and 2006).

8. In the British Columbia commercial halibut ITQ program, regulators froze trades for two years to allow participants to gain some understanding of how the system would work. Vessel owners thus retained their pre-ITQ allocations and made changes in fishing practices that reduced costs and raised revenues associated with their allocations.

9. For a discussion of GHL as used in the Pacific halibut sport fishery, see Criddle (this volume).

References

Abbott, Joshua, and James E. Wilen. 2008. *Rent Dissipation in Chartered Recreational Fishing: Inside the Black Box.* Draft manuscript.
Abbott, Joshua, Vishwanie Marahaj, and James E. Wilen. 2008. *Designing ITQ Programs for Commercial Recreational Fishing.* Draft manuscript.
Baeth, John L. 2007. *Halibut News.* (Available at http://www.halibut.net/halibut-news.html.)
Bluemink, Elizabeth. 2005. Halibut Harvest Could Force Showdown with Fishermen, Council. *Juneau Empire,* September 11. (Available at http://www.juneauempire.com/stories/091105/sta_20050911003.shtml.)
Criddle, Keith R. 2008. Examining the Interface between Commercial Fishing and Sportfishing: A Property Rights Perspective. This volume.
Dugan, Darcy, Ginny Faye, and Steve Colt. 2007. *Nature-Based Tourism in Southeast Alaska: Results from 2005 and 2006 Field Study.* Anchorage: University of Alaska, Institute of Social and Economic Research (ISER) and Eco-Systems. March 20. (Available at http://www.iser.uaa.alaska.edu/Publications/SEnbt_final.pdf.)
Griffith, David R. 2007. The Ecological Implications of Individual Fishing Quotas and Harvest Cooperatives. *Frontiers in Ecology and the Environment.* (Available at www.Frontiersinecology.org.)
Herrmann, Mark, S. Todd Lee, Keith R. Criddle, and Charles Hamel. 2001. A Survey of Participants in the Lower and Central Cook Inlet Halibut and Salmon Sport Fisheries. *Alaska Fishery Research Bulletin* 8(2): 107–17.
Homans, Frances R., and James E. Wilen. 1997. A Model of Regulated Open Access Resource Use. *Journal of Environmental Economics and Management* 32(1): 1–21.
———. 2005. Markets and Rent Dissipation in Regulated Open Access Fisheries. *Journal of Environmental Economics and Management* 49(2): 381–404.
Johnston, Robert J., Daniel S. Holland, Vishwanie Maharaj, and Tammy Warner Campson. 2008. Fish Harvest Tags: An Attenuated Rights-Based Management Approach for Recreational Fisheries in the U.S. Gulf of Mexico. This volume.
Leal, Donald R., and J. Bishop Grewell. 1999. *Hunting for Habitat: A Practical Guide to State-Landowner Partnerships.* Bozeman, MT: PERC. (Available at http://www.perc.org/perc.php?id=119.)
Loy, Wesley. 2007. Halibut Wars Continue to Roil Alaska's Competing Fleets. *Pacific Fishing* February: 10.

Sharp, Basil M. H. 2005. ITQs and Beyond in New Zealand Fisheries. In *Evolving Property Rights in Marine Fisheries*, ed. Donald R. Leal. Lanham, MD: Rowman & Littlefield Publishers, 193–211.

U.S. Fish & Wildlife Service (USFWS), Wildlife & Sport Fish Restoration Program. *National Survey 15 Year Trend Information*. (Available at http://wsfrprograms.fws.gov/Subpages/NationalSurvey/15_year_trend.htm.)

Wilen, James E. 2005. Property Rights and the Texture of Rents in Fisheries. In *Evolving Property Rights in Marine Fisheries*, ed. Donald R. Leal. Lanham, MD: Rowman & Littlefield Publishers, 49–67.

———. 2006. Why Fisheries Management Fails: Treating Symptoms Rather than Causes. *Bulletin of Marine Science* 78: 529–46.

Part IV

MANAGEMENT STRATEGIES FOR SALTWATER ANGLERS

Chapter 8

Fish Harvest Tags: An Attenuated Rights-Based Management Approach for Recreational Fisheries in the U.S. Gulf of Mexico

Robert J. Johnston, Daniel S. Holland, Vishwanie Maharaj, and Tammy Warner Campson

Although historically recreational fisheries have been perceived as having minimal impacts on U.S. marine fish stocks, it is now clear that activities of the recreational sector can have significant impacts, particularly in areas such as the Gulf of Mexico (Coleman et al. 2004). The long-term economic value of recreational fisheries is threatened by recent regulatory trends giving the saltwater angler shorter fishing seasons, smaller daily (or trip) bag limits, and more restrictive size limits for retaining fish. Such trends are often indicative of an inability to reduce fishing mortality to sustainable levels. Compounding the problem is the tendency to impose homogeneous fishing regulations over large regions, often on heterogeneous angler populations, which can lead to angler dissatisfaction and further loss of economic value (Sutinen and Johnston, this volume).

In response to such problems, and in recognition of the emerging success of individual fishing quotas (IFQs) in commercial fisheries (e.g., NRC 1999; Grafton et al. 2006; Leal, De Alessi, and Baker 2006), there is increasing interest among policy makers, stakeholders, and researchers in the potential application of a similar rights-based approach to recreational fisheries.[1] Potential advantages include longer seasons, improved management flexibility and control over fish mortality, increased economic benefits, and enhanced conservation motives among fishery participants. However, implementation of rights-based management in the recreational sector can face unique challenges partly because the sportfishing public is not easily integrated into these systems. Sutinen and Johnston (this volume) discuss challenges of integration, including the allocation of permits among large numbers of heterogeneous anglers, the monitoring of recreational harvest over these anglers, and potential ethical concerns with selling or owning durable rights to recreational fishing. Such challenges suggest that alternative rights-based methods may be needed. One alternative involves the use of harvest tags, similar to those currently used to manage fishing and hunting programs in the United States and abroad. Though still uncommon in recreational fisheries, harvest tags may offer some of the benefits of IFQs and related approaches, while avoiding some of the most difficult challenges.

This chapter explores the potential application of a harvest tag system to the Gulf of Mexico (GOM) recreational reef fish fishery, with an emphasis on red snapper and grouper species. Included is a review of the performance and attributes of various fishing and hunting tag programs applied worldwide. Conceptual, theoretical, and practical issues surrounding the application of harvest tags are also addressed. This is followed by a discussion of the challenges and opportunities related to the design of recreational harvest tag programs for GOM recreational reef fish fisheries. Insights from this paper should be applicable to a variety of recreational fisheries that face high angler demand and limited availability of fish.

Attenuated Rights-Based Fisheries Management

The term "rights-based" in the fisheries literature is often used to denote a management approach that assigns individual harvest rights (or privileges) to eligible anglers. Scott (1988 and 2000) identifies four primary characteristics of property rights which determine their ability to achieve efficient outcomes and foster good resource stewardship. These are exclusivity, transferability, durability, and security. Exclusivity is the extent to which the holder can exclude others from using or interfering with the holder's use of the property. Transferability is the extent to which the rights holder is free to transfer the property. Durability is the length of time the rights holder may exercise the powers above. Security is the ability of the rights holder to withstand challenges by other individuals, organizations, or government, to maintain rights to the property.

Harvest tags function through the assignment of a right to a specified quantity and type of harvest during a specified time period to each hunter or angler. Practical experience indicates that a well monitored and enforced harvest tag program can serve as a direct and practical way of ensuring a sustainable harvest. However, these rights are generally good for only a single season and nontransferable upon their issuance to hunters and anglers. For game, the tags can also specify a variety of associated limitations, including the sex or maturity of the animal (e.g., buck versus doe deer tags and bull versus spike elk tags) that can be legally harvested with a given tag, and the geographic area of harvest. On a continuum of rights-based approaches from weak to strong harvest rights, the rights conferred by tags are weaker (or more attenuated) than those conferred by IFQs, which are typically of unlimited duration and transferable among anglers. Indeed, some might not consider harvest tags to be a true "rights-based" management approach. As one increases the transferability and durability of the rights conferred by harvest tags, however, they become more akin to the stronger rights associated with IFQs. Strong harvest rights are typically associated with more efficient outcomes (i.e., greater net economic benefits) and good stewardship, but weaker rights might be more easily adapted to large-scale recreational fisheries. This trade-off is fundamental to the design of any potential harvest tag program for GOM recreational fisheries.

Status and Trends in GOM Recreational Reef Fish Fisheries

Trends in the GOM recreational reef fish fishery show evidence that current management mechanisms are inadequate to maintain harvest at levels consistent with a sustainable fishery and the maximization of net benefits to anglers. Red snapper, for example, is currently classified by the National Marine Fisheries Service (NMFS) as both overfished and subject to overfishing. Since 1992, recreational landings have often exceeded the quota allocated to the recreational fishery. The Reef Fish Fishery Management Plan was implemented in November 1984 and imposed a minimum size limit of thirteen inches for red snapper with the provision that anglers could keep a total of five undersized fish and an unlimited number of fish that exceeded thirteen inches fork length (GMFMC 1984). Through other regulatory actions since 1984, the daily bag limit has been progressively reduced from seven to four fish, the minimum size limit for retaining fish has been progressively increased from thirteen to sixteen inches, and the recreational season has decreased from a year-round season to one running from April 21 to October 31 (GMFMC 1997a, 1997b, 2005). In addition, in 2002, a three-year moratorium was imposed on new permits for charter vessels and headboats in the reef fish fishery (GMFMC 2003).

As with red snapper, progressively more restrictive management measures have been applied to the harvest of groupers, particularly red grouper.[2] The most recent red grouper population assessment (completed in 2007) concluded that the Gulf of Mexico population is neither overfished nor subject to overfishing. This conclusion, however, is at odds with prior findings that the stock is both overfished and subject to overfishing, and is undergoing additional review (NMFS 2007a and 2007b). From an aggregate daily bag limit of five fish of any grouper species, the bag limit for red grouper has been progressively diminished to two fish, with a temporary rule effective August 2005 further reducing this bag limit to one fish (NMFS 2006), and a final rule effective July 17, 2006, that established a permanent one red grouper limit per person per day (NOAA 2006). The minimum size limit for landing fish has also increased from eighteen to twenty inches (GMFMC 2005).

Management reform to a rights-based regime that incorporates elements such as fish tags in GOM recreational fisheries will face certain challenges due to a number of the fisheries' attributes, including (1) a large number of anglers from a wide geographic region; (2) highly heterogeneous user groups, including local and nonlocal private anglers and a large for-hire sector; (3) the lack of a small number of easily observed landing points at which the recreational sector—particularly private anglers—lands fish; and (4) anglers habituated to combinations of season, size, and bag limits, but also to the ability to fish at any time during the legal season, regardless of prior planning or arrangements other than obtaining a state saltwater license.

Harvest Tags in Hunting and Fishing

Hunting, like fishing, is often subject to significant restrictions to sustain scarce wildlife populations.[3] Beginning from a tradition of open access (subject to land access rights), all U.S. states now regulate hunting to some degree, and many have enacted complex combinations of management measures (e.g., hunting times; hunting licenses, stamps, and tags; and gear restrictions) to manage the harvest of various species of game.[4] In contrast to large-scale recreational fisheries with limited control over fish mortality and trends toward more restrictive management, however, managers in many states report successful and sustainable management of terrestrial game species.[5]

Among the more common measures used to control fowl and big game (e.g., deer, elk, and moose) hunting in the United States is the harvest tag. Harvest tags are issued by state wildlife agencies and authorize the taking of a specified number of animals from a designated species, often in a specific area or zone during the hunting season. They are typically applied in conjunction with other restrictions such as restricted hunting times (e.g., daylight hours) and methods of hunting (e.g., minimum caliber rifle) during the season. As most hunting tag programs share these and other common features (e.g., harvest quotas are established through limits on tag numbers; tags are distributed through officially approved channels), a reasonable understanding of tag programs may be obtained by examining a sample of existing programs. Table 8.1 summarizes primary attributes of a sample of state hunting tag (or mixed hunting and fishing tag) programs reviewed for this chapter.

Table 8.1 also provides details of the U.S. Migratory Bird Hunting and Conservation Stamp Program, commonly known as the Federal Duck Stamp Program (USFWS 2002). This program requires individuals over the age of sixteen to purchase a Duck Stamp in order to hunt various species of waterfowl in the United States.[6] It differs in many respects from the state hunting tag programs discussed here. For example, the primary goals of the Federal Duck Stamp Program are the generation of revenue for conservation purposes and the collection of data. Harvest restrictions are imposed at the state level, using traditional methods such as seasons and daily bag limits. Nonetheless, the Federal Duck Stamp Program is included here to illustrate the variety of tag programs in place nationwide, including those that do not directly limit harvest but have alternative goals.[7]

The attributes of hunting tag programs vary depending on factors including the scarcity of the species relative to hunting demand, the proportion of state resident versus nonresident participants, the purpose of the program, complementary regulations that may be in place, and other factors. The primary goals of hunting tag programs include (1) limiting harvest, (2) ensuring equitable distribution of harvest opportunity, (3) promoting effective monitoring and enforcement, and (4) providing data to improve management. Tag programs can either be simple or complex, with different classes of tags issued to different types of hunters (e.g., resident versus nonresident hunters or hunters using different types of equipment), using different allocation mechanisms (e.g., lottery, direct sales, and auctions), and with different types of approaches to ensure distributional fairness.

In general, hunting tag programs may be categorized in terms of their complexity and restrictiveness, which, in turn, often depend on the demand for hunting a game species relative to the number of animals that may be harvested. For example, states such as Connecticut and Pennsylvania incorporate hunting regulations of moderate complexity and restrictiveness, in which regulators set specific harvest quotas for game such as white-tailed deer in different regions, but in which the sustainable harvest in most regions exceeds hunters' demand (CT DEP 2006; PGC 2006). The number and restrictiveness of tags issued by such programs often varies by type of animal and area, given that in some areas a primary goal is the reduction of growing wildlife populations while in others it is the limitation of hunter harvest (e.g., CT DEP 2005). The use of harvest tags is common in such states, and most tags are available without strict rationing. Licenses are allocated regionally and often sold on

Table 8.1
Characteristics of Selected Hunting Harvest Tag Programs and Federal Duck Stamp Program

Location / Species	Allocation Method for Oversubscribed Hunts[a]	Tag Cost[b] (Adult Residents)	Number of Tags for Select Species[c]
Idaho			
all big game	lottery with preference points	deer: $20 elk: $30	general season deer: unrestricted controlled deer: 14,824 (2004) elk: 20,254 (2004)
Maine			
moose any deer bonus deer	lottery with preference points (moose only)	moose: $52 any deer: free bonus deer: $13	moose: 2,825 (2006) any deer: 67,725 (2006)
Montana			
all big game	lottery with bonus point system; bonus points cost $2 for residents	deer: $16 elk: $120	mule deer: 207,330 (2003) elk: 139,914 (2003)
Nevada			
all big game	lottery with bonus points	deer: $30 elk: $20	mule deer: 10,357 (2005) antlered elk: 843 (2005)
Oregon			
all big game	lottery with preference points	deer: $19.50 elk: $34.50	buck deer: 66,852 (2005) elk: 48,822 (2005)

Table 8.1 Continued
Characteristics of Selected Hunting Harvest Tag Programs and Federal Duck Stamp Program

Location / Species	Allocation Method for Oversubscribed Hunts[a]	Tag Cost[b] (Adult Residents)		Number of Tags for Select Species[c]	
Wyoming					
all big game	lottery with optional preference points[d]	deer:	$35	mule deer:	81,933 (2004)
		elk:	$47	elk:	58,852 (2004)
Colorado					
all big game	lottery with preference points	deer:	$31	deer:	119,593 (2004)
		elk:	$46	elk:	164,766 (2004)
Florida					
deer	permits distributed per lottery	deer:	varies	special deer:	118 (2004)
				special deer:	118 (2005)
Federal					
Duck Stamp	general sales		$15		1,600,000

Notes: All programs require physical tags except for the Florida deer and Federal Duck Stamp programs. Idaho, Maine, and Nevada have mandatory harvest reporting.
 a. Oversubscribed hunts are those for which hunter demand for tags exceeds the number available. For hunts that are not oversubscribed, multiple point of sale and direct distribution mechanisms are used in different states.
 b. Except for Florida deer and migratory birds, tags create an annual limit on individuals and limit total harvest.
 c. Year for which data is available on the number of tags distributed to hunters is recorded in parentheses.
 d. The cost of points for the Wyoming big game hunting program: $50 for elk and $40 for deer.
Source: Johnston et al. (2008, 16).

a first-come, first-served basis until the season is over or until quotas on tag numbers, where applicable, are met.

The Federal Duck Stamp Program could also be considered a program of moderate complexity, although it imposes no direct quota or restriction on harvest. Hunters must purchase a Federal Duck Stamp in order to be able to hunt migratory birds, but seasonal bag limits are not rationed at an individual level (USFWS n.d.; Freese and Trauger 2000). In contrast to many state hunting tag programs, the primary purpose of the Federal Duck Stamp Program is revenue generation, with 98 percent of tag revenues used to purchase wetlands and wildlife habitat for conservation purposes (USFWS 2002).

For species or regions in which demand for hunting exceeds a sustainable harvest, management agencies have implemented a variety of more intensive rationing systems for allocating hunting tags.[8] These include:

(1) *Limited harvest with lottery rationing.* In regions where demand for hunting exceeds the supply of animals available for a sustainable harvest, licenses or tags are often distributed through a lottery. Examples include rationing of deer tags or harvest rights in Maine, Idaho, and areas of Florida. Lotteries often distinguish between resident and nonresident hunter applicants, with more tags allocated to residents.

(2) *Limited harvest with "enhanced" lottery rationing.* To enhance the likelihood that repeat applicants who were unsuccessful in prior lotteries will be rewarded with hard-to-obtain tags and to ensure that tag allocation is perceived as equitable, some states use preference or bonus point systems. Examples are found in Colorado, Nevada, Oregon, Montana, and Maine. If a hunter enters a drawing and is not picked, she earns a point which may either serve as an additional chance in the next year's draw (a bonus point) or may place her into a limited pool of applicants (a preference point).

(3) *Auction of hunting rights.* Some states generate revenue either by auctioning a limited number of hard-to-obtain tags (e.g., moose in Maine) or by holding a special lottery in which hunters may purchase an unlimited number of chances to obtain desired tags. Given equity concerns, states usually allocate only a small percentage of available tags using such methods. Resulting revenues typically fund wildlife management activities.

Complementing primary harvest tag programs such as those reviewed in table 8.1, some states have also implemented tag or permit programs in coordination with private land hunting and habitat programs. For example, California has enacted a program whereby landowners can sell tags or permits directly to hunters. In addition to providing landowner economic incentives for providing wildlife habitat, these programs control the take on private lands through the issuance of tags or permits allocated to individual landowners. Such programs are specifically designed to generate economic incentives for private conservation, ensure area-specific data collection and management, and provide a localized, flexible approach to harvest tag allocation geared to the habitat and wildlife conditions of particular properties. The number of tags allocated to landowners is determined on the basis of the quality of wildlife and habitat on individual properties. Participating landowners are free to sell tags at market prices (although variations exist, for example a similar program in Utah includes a lottery to allocate a proportion of permits to public hunters). These programs are discussed in more detail by Freese and Trauger (2000) and Leal and Grewell (1999).

In contrast to many hunting tag programs, the majority of fisheries harvest tag programs exist primarily to improve information on catch and effort. Several programs, however, are used to control harvest. Most programs are less than ten years old. Table 8.2 summarizes eight of the better known harvest tag programs for recreational fisheries. These include (1) the pink snapper fishery in the Freycinet Estuary in Western Australia; (2) the recreational paddlefish fishery in the Missouri River below Gavins Point Dam in South Dakota; (3) the salmon and sea trout fishery in Ireland; (4) the recreational food-fish program for cod in Newfoundland; (5) the recreational tarpon fishery in Florida; (6) the recreational billfish fishery in Maryland and North Carolina; (7) the multispecies programs for recreational catch of salmon, steelhead, halibut and sturgeon in Oregon; and (8) the multispecies record card program in Washington State.

This review addresses only those programs that cover all recreational catch of the species being managed. It excludes programs that use tags to manage catch or limit landings outside of specified fish retention sizes, also called slot limits (e.g., the program for red drum in Texas) or to allow landings in excess of normal bag limits (e.g., the program for striped bass in New Jersey). Moreover, some states maintain similar programs for identical species that cross state borders, such as the paddlefish tag programs in Nebraska and South Dakota. For such similar programs, we review only one for the purposes of this chapter.[9]

As with the hunting tag programs above, all reviewed fishing tag programs are motivated by concerns over the sustainability of harvests. However, unlike hunting programs, relatively few fishing programs (e.g., the exceptions are pink snapper, paddlefish and tarpon) use tags to institute hard overall harvest caps. All reviewed programs except Florida tarpon and Maryland and North Carolina billfish, however, set limits on the number of fish an individual can

Table 8.2
Characteristics of Reviewed Harvest Tag Programs for Recreational Fisheries

Location Tag Type / Allocation	Species	Cost of Tag (adult resident)	Tags Create Limit on Catch Individual / Total Catch	Number of Tags Sold[a] (per year) / Oversubscribed	Mandatory Harvest Reporting
Western Australia, Shark Bay attach / lottery	pink snapper	AUS$10	✓	1,400 ✓	
Ireland attach / with license	salmon and sea trout	free with license	✓	~25,000	✓
Newfoundland attach / with license	cod	free with license	✓	~135,000	✓
South Dakota, Missouri River attach / lottery	paddlefish	US$5	✓	archery: 275 ✓ snagging: 1,400 ✓	
Florida attach / purchase	tarpon	US$51.50	✓[b]	300–400	✓
North Carolina and Maryland attach / available at designated sites	bluefin tuna, white & blue marlin, sailfish, swordfish	free		~2,000–3,000	

Table 8.2 Continued
Characteristics of Reviewed Harvest Tag Programs for Recreational Fisheries

Location Tag Type / Allocation	Species	Cost of Tag (adult resident)	Tags Create Limit on Catch Individual / Total Catch	Number of Tags Sold[a] (per year) / Oversubscribed	Mandatory Harvest Reporting
Oregon record card / in addition to license	salmon, steelhead, halibut sturgeon	US$21.50	✓	208,452	
Washington record card / in addition to license	salmon, steelhead, halibut sturgeon, Dungeness crab	card free with license	✓	~650,000	✓

a. Except for Florida, Australia, and Ireland, there are no limits on the number of tags that can be issued. Refer to Johnston et al. (2008) for details on the number of tags sold or the estimated number of tags sold per year.
b. Even though the number of tarpon tags is capped at 2,500, there is no binding limit on total catch.
Source: Johnston et al. (2008, 27).w

land annually. These limits are likely not binding for the majority of anglers who fish moderately during the season but constrain catches by some and, therefore, contribute to limiting total catch. None of the programs relies solely on tags to manage catch. Size limits and either daily bag limits or season restrictions (or both) are in place in most cases. In some cases, however, tag program managers explicitly note that the existence of harvest tags has allowed relaxation of previous management measures. For example, the South Dakota paddlefish tag program has allowed for a longer open season (Stone and Sorenson 2002).

Most programs distribute tags that must be attached to a fish as soon as it is caught and retained. The program for various species of billfish in coastal waters off Maryland and North Carolina, in contrast, does not require anglers to possess or affix a tag when the fish is caught, only when landed.[10] The Washington catch record card and the Oregon combined harvest tag do not use physical tags. Rather they provide a booklet in which anglers must record catches and locations of all fish retained.[11]

Like the hunting tag programs discussed above, the number of allocated tags in fish tag programs depends on program goals and other factors. For example, the total number of paddlefish tags allocated each season is greater than the overall catch that managers target for the season, based on historical patterns showing that only about half of the tags will be used to retain fish.[12] In contrast, although only about half of historically allocated pink snapper tags are used to retain fish, the number of allocated tags is maintained at the targeted overall catch level to ensure that level is not exceeded.[13] Tarpon tags in Florida are capped at 2,500, but only about 300 to 400 are actually issued to anglers each year because retaining tarpon is discouraged by managers, the cost of tags is relatively high, and retaining catch is not usually desirable due to the low food value of the fish.[14] The Irish salmon tag program has a targeted recreational catch of about 15,000 fish per season, but the total number of tags is not limited as each salmon license is allocated twenty tags and the number of licenses is not capped.[15]

Fishing tag programs use a variety of means to address the situation in which the demand for tags is greater than the supply. The pink snapper and paddlefish programs use lotteries to ration tags, much like many hunting tag programs (Mestl 2001).[16] In the case of paddlefish, there are separate lotteries for archery and snagging methods as well as for state residents and nonresidents. In the pink snapper fishery, the purchase of tags is not limited to anglers, and environmental groups have applied for tags to reduce harvest.[17] The tag program for Florida tarpon sets a relatively high price for tags (US$51.50) that discourages acquisition, keeping the quantity demanded well below the maximum available.[18] Other programs, such as those in Oregon, have lower tag fees that may have some impact on the quantity of tags demanded.[19] Notably, none of the programs allows resale of tags, although pink snapper tags, once obtained, may be given away by anglers.[20]

In all reviewed cases, managers report that harvest tag programs have fully or partially met their objectives. For example, the pink snapper program has maintained actual overall catch each season below target levels.[21] The paddlefish program has maintained overall catch at desired levels, reduced crowding in popular areas, and allowed a longer season (Stone and Sorenson 2002).[22] The tarpon tag program has been successful at reducing retained fish but has proven less effective at estimating harvest rates due to incomplete reporting and the use of some tags for temporary possession of fish while they are being weighed and then later released live.[23] The Newfoundland cod program has improved management by providing enhanced data collection on catch and effort as well as by raising public awareness of concerns over conservation of groundfish stocks.[24] The Irish salmon tag program provides data on catch levels and exploitation patterns which are used to guide management decisions.[25]

Despite these reported successes, most programs have also faced some challenges. Educating the public and generating stakeholder support has been a particular concern. Managers for pink snapper, Irish salmon, Newfoundland cod, and Oregon and Washington multispecies fisheries report resistance by some anglers related to the cost of tags, the inconvenience of using them, or other aspects of the program. In general, however, most programs are well received.[26]

In sum, both fishing and hunting harvest tag programs have shown success in managing the recreational harvest of valued wild stocks. The attenuated rights-based characteristics of tag programs have enabled many to maintain harvest below target levels, while, in some cases, allowing the relaxation of traditional restrictions and negative trends in management, such as the progressively shorter seasons and smaller bag limits found in the Gulf of Mexico red snapper and grouper fisheries. In addition, most programs have been generally (although not universally) well received by both anglers and managers.

The Potential of Harvest Tags in GOM Recreational Fisheries

As noted above, implementation of harvest tags varies depending on the attributes and context of the harvested resource(s) in question. Challenges and opportunities related to the use of harvest tags in the GOM recreational reef fish fishery must be considered within the context of the fishery, angler attributes, and potential goals of management. The review of hunting and fishing tags summarized above highlights seven potential advantages of harvest

Table 8.3
Harvest Tags vs. Current Management in Gulf of Mexico Recreational Fisheries

Issue	Features of Current Management	Features of Harvest Tags
Harvest Limits	No effective harvest limits imposed; quotas are "soft" or do not exist. Trends toward more restrictive management.	Allows hard harvest limits to be imposed. Would require large number of tags and complex administration. Number of tags should account for potential release mortality.
Season Length	Trends toward shorter seasons related to ineffective harvest control.	Allows for longer seasons compared to nonrights based management promoting angler satisfaction.
Rights Allocation	Harvest open to all anglers subject to license, bag, size, season, and other limits. Rights allocation not a concern, as management is not rights based. Waiting period or pre-planning rarely required to fish.	Requires establishment of mechanisms for allocation of tags. Allocation complicated by large number of anglers, heterogeneous groups, and resident vs. nonresident distinctions. Short-term rights can ameliorate allocation concerns. Allocation methods for scarce tags include lotteries and auctions. Examples of various successful allocation modes in existing programs. May involve monetary costs and waiting periods to obtain tags.
Monitoring, Enforcement, and Compliance	Faces challenges associated with monitoring, enforcement, and compliance with regulations in large scale fisheries.	Monitoring challenging but ameliorated by attributes of tags (such as observability at check points). Can increase voluntary compliance and self-policing among anglers. Angler education and information materials often required.
Data Collection	Recent assessments identify limitations with current methods of obtaining data for fisheries, including those in GOM.	Tags can provide data on some or all aspects of recreational fishing. Reporting and data gathering in tag programs provides lessons for GOM fisheries. Reporting compliance varies with incentives provided by program.

Table 8.3 Continued
Harvest Tags vs. Current Management in Gulf of Mexico Recreational Fisheries

Issue	Features of Current Management	Features of Harvest Tags
Revenue Generation	Bag and size limits provide no mechanism for cost recovery or revenue generation.	Revenues from the sale of tags can support management, education, and data collection. Revenues must be viewed within the context of program cost.
Sector Integration	Private and for-hire sub-sectors face same bag, size, and season limits. Regulations rarely suit groups equally. Commercial and recreational sector not integrated under current management.	Many models for integration of management for private and for-hire groups using harvest tags programs. Possibility of rights transfer between recreational and commercial sectors; practical mechanisms for integration are not well developed.

Source: Johnston et al. (2008, 29).

tags for GOM recreational fisheries. These include (1) the ability to maintain harvest within hard harvest limits; (2) the potential for longer seasons; (3) the availability of mechanisms to promote equitable tag allocation; (4) the ability to contribute to more effective monitoring, enforcement and compliance; (5) the provision of more extensive harvest data; (6) the generation of revenue; and (7) the potential ability to integrate with for-hire recreational and commercial fishery sectors. As summarized in table 8.3, each of these potential advantages, however, comes with the potential for concomitant challenges or disadvantages.

Hard Harvest Limits

GOM recreational fisheries for snapper and grouper account for a substantial and rising proportion of total fishing mortality, so failure to constrain catches within allocated quotas presents a threat to sustainability (Coleman et al. 2004). Current management methods in GOM recreational fisheries, which include size limits, daily bag limits, and restricted seasons, do not impose hard harvest limits. In contrast, harvest tags combined with existing harvest quotas for many Gulf species provide a potential mechanism to introduce hard harvest limits into recreational fisheries. There are numerous cases of state hunting programs in which harvest tags are used to impose such limits. While the use of tags to impose hard limits is still rare in fishing applications, the fundamental mechanism for linking tag management to such limits is well established. Generally, harvest targets for specific species are established through a process involving policy makers, scientists and sometimes stakeholders. These targets may be established over small or large geographical regions, depending on management goals. Tags are then issued allowing a harvest quantity equal to this quota. In some cases, the number of tags issued may be larger than the quota based on estimates of the number of tags that will go unused.

Recreational quota has already been established for red snapper at 4.47 million pounds (GMFMC 2003), and a recreational target catch level for red grouper has been established at 1.25 million pounds (NOAA 2004; NMFS 2007a). For these species, harvest tags would provide a mechanism to enforce current quotas and/or targets. For species with no recreational quota, such as gag grouper, the implementation of harvest tags would require the establishment of a recreational quota. In all cases, an additional step would be the translation of current quotas, which are specified in pounds, to those suitable for recreational harvest tags, which are specified in number of animals and fish.

One concern is the potentially large number of tags corresponding to the harvest limits in the GOM reef fish fishery. Most hunting applications involve fewer than 250,000 tags, with the majority of programs issuing fewer. Most fish tag programs involve similar numbers of tags. An exception is the Washington State program, in which approximately 650,000 catch record cards for salmon, steelhead, sturgeon, halibut, and Dungeness crab are issued annually.[27] In contrast, tag programs for species such as GOM red snapper would likely require roughly one to two million tags, assuming one tag per retained fish, and would involve up to five states' waters and federal waters. Experience from the program in Washington suggests that tag programs can incorporate and track large numbers of tags. However, one might also expect that large tag numbers would be associated with a larger required administrative structure and perhaps program cost.

An additional issue in determining tag numbers and harvest limits is pre- and post-release mortality. The impacts of release mortality are an important consideration in most forms of recreational fishery management, and recent research suggests that release mortality in recreational fisheries may be higher than was previously assumed (e.g., Schirripa and Legault 1999; Millard et al. 2003; Woodward and Griffin 2003). The use of harvest tags would likely continue to require the release of fish for which harvest is not permitted. Modified tag systems might, however, be designed to mitigate some of the problems leading to increased discards and related mortality. For example, a tag system might ameliorate the likelihood of high-grading (i.e., discarding smaller fish in order to keep fishing for larger ones) through the creation of varying tag classes for different sizes of fish, much as tag programs for deer often require different tags for does versus bucks. Anglers, for example, could keep smaller fish without using their regular fish tags, thereby reducing the incentive for wasted throwbacks of these smaller fish. This would be similar to current systems that require tags to retain fish outside of slot limits (e.g., Texas red drum). In addition, more effective fishery management and resulting stock increases (for example, though a tag system) might, over time, allow for greater levels of retained recreational harvest, thereby reducing the necessity to discard fish for which retention is not permitted.

Season Length

While there may be some disadvantages to longer open seasons (e.g., greater costs of monitoring and enforcement), reductions in season length generally diminish economic benefits derived from recreational fisheries, are a symptom of ineffective long-term control of harvest mortality, and are viewed by many as revealing the "fundamental flaws" of command-and-control fisheries management (Sutinen and Johnston, this volume). Currently, the GOM red snapper season runs from April 15 to October 31, down from a year-round season in place through 1996. While the grouper season is currently year-round, it is anticipated that there will be a seasonal grouper closure from February 15 to March 15 beginning in 2007 (GMFMC 2005). Current management methods will not lead to a lengthening of open seasons. In contrast, harvest tag programs, such as the South Dakota paddlefish tag program, have resulted in a lengthening of fishing seasons.[28] Other hunting tag programs reveal success in maintaining harvest limits and sustainable species populations, with no evidence of trends toward reduced season lengths (Johnston et al. 2008). The potential for longer open seasons is one of the more significant benefits associated with rights-based management, including attenuated rights-based systems such as harvest tags.

Allocation of Fishing Rights

One of the most difficult issues facing the implementation of IFQs in commercial fisheries is the allocation of fishing rights—a challenge not faced under traditional recreational management mechanisms such as bag and size limits. Hunting and fishing tag programs incorporate an array of provisions to promote equitable allocation of tags. Mechanisms include (1) lotteries for high-demand tags, including preference or bonus points or both; (2) tag set-asides for particular groups or harvest methods; and (3) limits on the number of tags that may be held by individuals. Less scarce tags (relative to demand) are distributed using a variety of point of sale and other mechanisms, including direct sales at fishing or hunting supply shops, sales through for-hire operators, and distribution with fishing licenses.

Tag allocation mechanisms range in complexity from simple programs such as the Newfoundland Cod Food Fishery in which tags are automatically distributed with groundfish licenses to complex systems such as those for Colorado deer and elk or Maine moose tags in which multiple modes of tag allocation are combined (e.g., lotteries, bonus point systems, and auctions) to address severe oversubscription (i.e., many more tags are demanded than are available). Most programs are of modest complexity. For example, the fishing tag programs for Western Australia (Shark Bay) pink snapper and South Dakota paddlefish rely on a straightforward lottery allocation. In general, programs in which the supply of tags are well short of demand are associated with more complex allocation mechanisms such as lotteries with preference points or bonuses or auctions, while programs in which the supply of tags exceeds demand are associated with less complex allocation mechanisms such as automatic allocation with the purchase of a license at sporting goods outlets.

Individual allocation methods have advantages and disadvantages. For example, market or price-based rationing (including auctions) can represent an effective means to allocate tags to those with the highest willingness to pay, but they can also result in a disenfranchisement of anglers with moderate to low levels of wealth. Lottery systems divorce allocation likelihood from willingness (and ability) to pay but provide no mechanism through which preference is given to those for whom tags are valued highly. First-come, first-served mechanisms can result in a race for permits and the unavailability of permits late in the season. Provision of tags with license purchases offers an allocation mechanism which can take advantage of existing systems for license sales; however, it does not control harvest unless the number of licenses is limited.

Related to the rationing issue is any waiting period that might be required to obtain a tag and potential uncertainty related to lottery allocation. Current regulations in the GOM recreational reef fish fishery allow fishing during any day of the open season, with no extensive prior planning. The ability to fish within the open season is also certain, at least with regard to regulatory constraints. In contrast, some types of tag distribution mechanisms would require a delay or waiting period between when an angler ordered a tag and when harvest could occur. Alternatively, lottery allocation would involve uncertainty as to whether tags could be obtained. Significant waiting periods or uncertainty could erode angler support for tag management. Distribution of tags through retailers (e.g., sporting goods or tackle and bait outlets) might alleviate that problem, but this might not be possible if a lottery system is used to ration tags.

Given the large number and heterogeneity of anglers fishing from private and for-hire boats, multiple-mode allocation may be the most suitable for meeting demand. Such allocation mechanisms are common, particularly for states with both under- and oversubscribed hunts, and are applied in many state hunting tag programs including those in table 8.1.[29] Options include allocation of tags with fishing licenses, with additional tags available through point of sale purchase, lottery, auction, or other mechanisms. One might also develop a mechanism whereby a certain proportion of tags are distributed to operators in the for-hire sector. Such provisions are common in hunting tag programs (e.g., Oregon and Idaho) and can help to ameliorate concerns that insufficient tags would be made available to support anglers wishing to use for-hire services (IFG 2006; ODFW 2005).

An additional and related issue is the allocation to residents versus nonresidents. Most tag programs distinguish between state resident and nonresident applicants, with the majority of tags available to residents. However, most harvests in the GOM reef fish fishery occur in waters of the federal Exclusive Economic Zone (EEZ), the area typically between three and two hundred miles offshore. Therefore, a distinction between Gulf state residents and nonresidents may not be relevant.

A final issue relates to tag transferability. Few traditional tag programs allow for the transfer of harvest tags, although some allow tags to be transferred or swapped contingent upon no money changing hands. One of the notable exceptions is the recently enacted Kansas transferable deer hunting permit program, which allows landowners to obtain (through a lottery) and transfer a specified quantity of hunting permits. Permits may be transferred "to any person with or without compensation" (Taylor and Marsh 2003, 4). Maine, Idaho, and Florida, among other states, allow more limited transferability of hunting tags; for example, tags may be transferred free-of-charge but not sold (IFG 2006; FWC 2005).[30] Transferability has both positive and negative aspects. Transferability allows tags to reach those for whom harvest is most highly valued, thereby maximizing the net economic benefits. However, permit or tag sales could also result in speculation for highly valued tags and possible "pricing out" of less wealthy anglers and might also increase the number of tags used and hence fish retained. The Kansas transferable deer permit program provides some evidence of this, with substantial differences between the cost of permits obtained directly through government lotteries and those purchased through landowners or guides (Taylor and Marsh 2003; Kim, Woodward, and Griffin, this volume).

Monitoring, Enforcement, and Compliance

Recreational fishery management often faces significant challenges associated with monitoring, enforcement, and angler compliance with regulations (Sutinen 1993). Given the large number of participants involved and the absence of central locations at which all participants may be intercepted and observed, recreational management in the GOM reef fish complex relies heavily on voluntary angler compliance, coupled with monitoring and enforcement measures. Harvest tags do not eliminate difficulties with monitoring, enforcement, and compliance. However, there are aspects of tag programs which can ameliorate some challenges. For example, a requirement that physical tags be attached to harvested fish together with random checks or checkpoints can facilitate monitoring and enforcement. Harvest reporting requirements associated with tags can also aid in more accurate harvest monitoring, particularly if anglers must report the number of tags actually used in order to obtain additional tags or tags in subsequent years.

Another area in which harvest tags could encourage improved monitoring and enforcement is in the form of voluntary compliance and self-policing behavior. Although not a panacea for all potential stewardship concerns associated with fishery management, rights-based arrangements in commercial fisheries are often associated with increased motives for stewardship and regulatory compliance (NRC 1999; Grafton et al. 2006; Sutinen and Johnston, this volume). Similar stewardship could be associated with short-term rights such as those granted by harvest tags (Macinko and Bromley 2002). While the harvest rights conferred by tags might encourage greater stewardship among anglers compared to current management, incentives for stewardship are likely to be less pronounced than those observed for durable rights-based programs such as IFQs in commercial fisheries. Beyond compliance related to stewardship motives, anglers who have made an outlay, in money or time, to obtain a fishing tag, may also have a greater incentive to report others who are harvesting illegally, compared to incentives that exist under current bag and size limits.

Harvest tag programs may also contribute to voluntary compliance by making anglers aware of resource scarcity, or by increasing anglers' satisfaction with management.[31] However, experience in the small number of fish tag programs is not universal in this regard. For example, the Newfoundland Cod Food Fishery has experienced a protest fishery (noncompliance with regulations as a form of protest) due to anglers dissatisfied with the program.[32] Discussions with program managers suggest that education programs can be critical to encouraging support for harvest tag programs.[33] For this reason, many tag programs incorporate education mechanisms. For example, Colorado provides a "planning tips for nonresidents" web page with links to regulations for hunting in the state.[34] Similar websites are available in other states, with the purpose of educating residents and nonresidents regarding hunting tag and license programs.

Providing Harvest Data

Recent assessments have identified significant limitations with current methods of obtaining data for recreational fisheries (CRRFSM 2006). A lack of accurate data can hamper development of appropriate management responses but also points to the potential value of management mechanisms such as harvest tags that can provide data on some or all aspects of recreational fishing. The wide array of harvest reporting and data gathering mechanisms incorporated into hunting and fishing tag programs—together with the relative successes and failures of these mechanisms—provides lessons for developing methods for GOM recreational fisheries. Voluntary compliance with harvest reporting, and hence data collection effectiveness, varies across programs. As one might expect, compliance with harvest reporting is greater in programs that impose negative consequences for noncompliance (Johnston et al. 2008). A requirement to provide data on tag use before acquiring another tag or license is the most common method to increase compliance. For example,

Idaho hunters are unable to obtain hunting licenses for the next year unless a harvest report card has been submitted for the current year (IFG 2006).

Revenue Generation

Current management mechanisms such as bag and size limits provide no mechanism for cost recovery. Revenues from the sale or auction of harvest tags, in contrast, can be used to support management, education, data collection, and other efforts. For example, between 1934 and 1996, the U.S. Migratory Bird Hunting and Conservation Stamp Program (Federal Duck Stamp Program) generated over US$442 million (Freese and Trauger 2000). Most all programs charge at least a nominal fee for tags (e.g., see tables 8.1 and 8.2). The resulting revenues, however, must be viewed within the context of the cost of implementing the programs. Many hunting tag programs, and all except the Florida Tarpon Tag program in recreational fisheries, require only nominal payments for tags; payments of US$5 to US$20 are common (table 8.2). As a result, revenues are often insufficient to cover the full cost of fishery management, although in some applications tag-generated revenue covered the cost of harvest tag administration. In general, there is a trade-off between a program's ability to raise funds using high tag prices and its ability to promote equitable tag distribution and widespread angler support.

For-Hire and Commercial Sector Integration

The GOM recreational reef fish fishery involves a large number of anglers who fish from private boats as well as those who use the services of the for-hire sector. Differences between these groups can lead to difficulties in developing regulatory mechanisms that apply to both. Moreover, the net economic benefits of fisheries can increase when rights-based management integrates all sectors of a fishery. Integrated rights-based management of all sectors with hard output controls can provide a transparent mechanism for reallocation of scarce fish to sectors with higher marginal values.

There are numerous models for integration of management for private and for-hire groups using harvest tag programs, many drawn from hunting contexts involving a large for-hire guide sector. Strategies for integration include (1) hunters can first obtain the desired tag, then seek out an appropriate for-hire guide or hunting service; or (2) guide services can obtain tags on behalf of hunters. To account for concerns about the availability of sufficient permits to support business operations, many states provide a set-aside allocation of tags for outfitters and guides or incorporate programs that guarantee for-hire hunting operations a certain hunt allocation. Similar models could be adapted for use in the GOM recreational reef fishery. For example, a certain percentage of harvest tags might be allocated to the for-hire sector in the reef fish fishery, to be made available to anglers who purchase fishing trips on charter or other for-hire vessels.

Tag programs also allow the possibility of harvest rights transfer between recreational and commercial sectors, following general examples such as those proposed for the Alaska halibut charter IFQ program (Criddle et al. 2003; Criddle, this volume). Practical mechanisms for such integration, however, are not well developed in either hunting or fishing applications of harvest tags, although a very small number of programs have allowed for the limited transfer of longer-term rights between commercial and recreational sectors (Arnason and Pearse, this volume).

Moreover, stakeholders can have philosophical, ethical, or utilitarian concerns regarding transfers of fishing rights between commercial and recreational sectors, as evidenced by concerns about potential transfers between the recreational and commercial sectors of the Irish salmon fishery,[35] as well as extended controversy regarding the implementation of a potential charter IFQ program for the Alaska halibut fishery, at least in part related to provisions that would allow limited transferability of quota shares between the commercial and recreational sectors (Smith and DeCosimo 2006). It is often unclear to what extent resistance to between-sector transfer is related to philosophical concerns per se (e.g., a belief that transfers to the commercial sector are illegal or unethical because they deny anglers the ability to utilize a public trust resource), versus rent-seeking (e.g., resistance related to a belief that transferability could lead to economic losses among certain groups). Regardless, past controversies related to recreational-commercial allocations (Gislason 2006), combined with resistance to programs that would allow limited transferability between sectors (Criddle et al. 2003; Criddle, this volume), suggest that any program that would allow tag-based transfer between sectors would have to be carefully designed to encourage acceptance among different stakeholder groups.

Coordination of State and Federal Programs

There is no federal saltwater angling license or any federal program that currently enables individualized communication with recreational anglers. Harvest tag management would require interaction with large numbers of anglers, if only to distribute tags and collect harvest data. Establishment of such a system is costly. This cost could be reduced, however, by coordination with state license programs. All Gulf states require a saltwater fishing license. Licenses are required for all saltwater anglers, with few exceptions.[36] These licenses are typically annual[37] and hence require yearly, individualized contact with the vast majority of resident and nonresident anglers who fish in state waters. The infrastructure and mechanisms for this state-level individualized angler contact provide a potential mechanism whereby tags and/or associated information could be distributed. For example, tags for certain species could be distributed along with state fishing licenses or could be purchased with licenses for an additional cost. Tags could be made available through any channel whereby licenses could be obtained, with states administering the program according to regional council guidelines.

Additional Variants of Tag Programs

One of the more common elements of tag programs is the either mandatory or voluntary bundling of tags for various species. Examples of voluntary bundling include the Oregon "Sports Pac," available only to residents, which includes a combination angling and hunting license as well as several tags including a combined angling and harvest tag, a deer tag, an elk tag, a bear tag, a cougar tag, a spring turkey tag, and validation for upland birds and waterfowl.[38] The similar Idaho "Sportsman's Package" includes tags for deer, elk, bear, mountain lion, turkey, salmon, and steelhead (IFG 2006). For these programs, one can purchase a separate species tag instead of the bundle. An example of mandatory bundling is the Oregon Combined Angling Harvest Tag, which may be purchased alone or as part of a Sports Pac.[39] This tag allows the angler to catch and keep twenty salmon and steelhead, five sturgeon, and six halibut. In this program, one cannot purchase an individual species tag. Advantages of tag bundling include ease of administration and reduced costs. A single tag-bundle is distributed for a group of species, rather than individually for each species. This can reduce administrative costs and burden, particularly for tags that are only sold as a bundle. Another advantage is that tag-bundles can provide anglers with the ability to retain fish for which harvest is not anticipated.[40]

Disadvantages of bundling (particularly mandatory) include problems with tag availability if different groups of anglers target different species. For example, if tags for red snapper were only available bundled with grouper tags, then grouper tags could be "sold out" through bundled purchase by red snapper anglers, even though these anglers have no intention of harvesting grouper. Anglers might also resist paying for unwanted tags bundled with desired tags. Such dissatisfaction has been associated with the Oregon bundling program.[41]

Another relevant issue of tag bundling is the potential for single tags that may be used to harvest fish of different species. For example, one might purchase a "grouper tag" that would allow the harvest of any one of a specified set of grouper species. Similar approaches apply to current bag limits in the grouper complex that allow for up to five grouper per day of a variety of (but not all) species. A combined tag might reduce administration cost and maximize the utility of single tag but would also reduce managers' ability to set harvest limits for specific species.

The flexibility of annual tag distribution mechanisms could allow for programs in which some or all of a selected set of harvest tags would be distributed to particular (perhaps disadvantaged) coastal communities—either for distribution to residents (or nonresidents) or for sale as a means to raise revenues. In effect, tags could allow for an intentional distribution of annual fishery rents to particular communities or other entities, as a means to accomplish various policy goals (e.g., community development).

Conclusion

Recent management trends suggest that regulatory mechanisms for recreational fisheries such as bag, size, and season limits are often unable to provide for sustainable fisheries and maximize the potential long-term benefits of the fishery to anglers. This review of existing harvest tag programs and analysis of potential application for the GOM recreational reef fisheries suggest that harvest tags represent a promising alternative or complement to the current system of bag, size, and season limits. Potential advantages of tags include increased control over total catch, improved information for management, and increased long-term benefits to anglers. Our review also highlights potential challenges associated with the development and implementation of harvest tags and suggests that realization of the advantages of harvest tags requires a well conceptualized plan tailored to the fisheries involved and to the preferences and attributes of anglers.

Given the potential complexity of successful harvest tag programs, challenges illustrated by existing programs, and the size of the fisheries in question, implementation of harvest tags for GOM recreational reef fish fisheries would likely require significant planning efforts at both the state and federal level. One means of testing the suitability and details of potential harvest tag programs for the recreational reef fish fishery would be through the use of small-scale pilot programs for individual species or small areas of the GOM, developed in coordination with anglers, managers, and other stakeholders. Managers would have to design a program suited to both the needs of stakeholders and biological fishery attributes. The design of such a program would require choices and trade-offs, and it would have to account for such factors as the size of the fisheries involved, the quantity of harvest consistent with a sustainable fishery, the heterogeneity of private anglers and for-hire operators, and the need to ensure equitable access to recreational fishing opportunities. The potential complexity of program design notwithstanding, the widespread success of harvest tag programs worldwide suggests that appropriately designed programs can result in sustainable harvest of fisheries and an increase in economic benefits relative to common recreational fishery management methods.

Notes

1. One example entails recent efforts by various fishing interests and the North Pacific Fishery Management Council (NPFMC) to implement an IFQ program in the charter-based, recreational sector of the Alaska halibut fishery. See NPFMC (2006) and Criddle (this volume).

2. There are also harvest bans for Goliath and Nassau grouper. Furthermore, a limit of one speckled hind per vessel and one warsaw grouper per vessel was included in the five grouper aggregate bag limit established in 1999. See GMFMC (1999 and 2005).

3. Given the emphasis of this paper on potential fisheries applications, we provide greater attention to the use of harvest tags in fisheries. As hunting tag programs are older and more developed, however, we also provide a broad summary of some of the key aspects of these programs as implemented in different states.

4. For example, a history of hunting regulations in Pennsylvania is provided by the PA Game Commission website (http://www.pgc.state.pa.us/pgc/cwp/view.asp?a=458&q=153947.)

5. Dan Holland conducted telephone interviews with W. Bolduc, Office of Public Information, Maine Department of Inland Fisheries and Wildlife, June 6, 2006; B. Compton, Idaho Department of Fish and Game, December 20, 2005 and December 22, 2005; T. Thornton, Oregon Department of Fish and Wildlife, January 3 and January 13, 2006; and E. Slater, Colorado Department of Wildlife, Limited Licensing Division, June 6, 2006.

6. Nonhunters such as stamp collectors and birders also purchase Federal Duck Stamps.

7. Additional information is provided by Johnston et al. (2008).

8. Because resident big game tags are typically sold at a nominal price set by state regulatory agencies, there is no market or price mechanism whereby excess demand can be eliminated. Hence, alternative mechanisms are needed to allocate tags in the presence of excess demand. For nonresident hunters, however, allocation can be through market prices. Such is the case in Montana where demand for nonresident big game licenses and tags far exceeds supply.

9. Additional details of these programs are provided in Johnston et al. (2008).

10. Dan Holland conducted telephone interviews with Ron Salz, Fisheries Statistics Division, U.S. National Marine Fisheries Service, April 12, 2006; and R. Dunn, Department of Sustainable Fisheries, U.S. National Marine Fisheries Service, April 18, 2006.

11. Dan Holland conducted telephone interviews with S. Markey, Acting Catch Record Card Project Manager, Washington Department of Fish & Wildlife, March 31, 2006; R. Messmer, Oregon Department of Fish and Wildlife, April 10, 2006; and Harry Upton, Oregon Department of Fish and Wildlife, April 27, 2006.

12. Dan Holland conducted a telephone interview with J. Sorenson, South Dakota Department of Game, Fish & Parks, Missouri River Fisheries, March 28, 2006.

13. Dan Holland conducted a telephone interview with N. Harrison, Program Manager Recreational Fisheries, Western Australia, Department of Fisheries, March 29, 2006.

14. Dan Holland conducted a telephone interview with J. Colvocoresses, Marine Fisheries Biology Subsection, Florida Fish and Wildlife Conservation Commission, April 27, 2006.

15. The number of tags allocated per license can be adjusted annually to influence total catch.

16. Telephone interview with J. Sorenson (see note 12).

17. Telephone interview with N. Harrison (see note 13).

18. Although lower than the price for many hunting tags, research shows that anglers can be fairly responsive to even relatively low prices charged for licenses or tags (Sutton, Stoll, and Ditton 2001).

19. Telephone interviews with R. Messmer (see note 11) and Harry Upton (see note 11).

20. Telephone interview with N. Harrison (see note 13).

21. Telephone interview with N. Harrison (see note 13).

22. Telephone interview with J. Sorenson (see note 12).

23. Telephone interview with J. Colvocoresses (see note 14).

24. Dan Holland conducted a telephone interview with B. Slade, Department of Fisheries & Oceans Staff Officer, Recreational Fisheries, April 26, 2006.

25. Dan Holland conducted a telephone interview with F. Grant, National Protection/Conservation Coordinator, Ireland Central Fisheries Board, March 31, 2006.

26. Telephone interviews with S. Markey (see note 11), Harry Upton (see note 11), N. Harrison (see note 13), B. Slade (see note 24), and F. Grant (see note 25).

27. Telephone interview with S. Markey (see note 11).

28. Telephone interview with J. Sorenson (see note 12).

29. See Johnston et al. (2008) for additional details.

30. Telephone interview with W. Bolduc (see note 5).

31. Telephone interview with J. Sorenson (see note 12).

32. Telephone interview with B. Slade (see note 24).

33. Telephone interview with S. Markey (see note 11).

34. See http://wildlife.state.co.us/Hunting/ResourcesTips/NonResidentTips.htm.

35. Telephone interview with F. Grant (see note 25).

36. For example, some states do not require licenses for minors or senior citizens.

37. An exception is the Lifetime License in Florida.

38. Telephone interview with Harry Upton (see note 11).

39. Telephone interview with Harry Upton (see note 11).

40. This may be important for species often caught together such as various species of grouper, but it might be less appropriate for more distinct fisheries. For example, it might be less appropriate to include red snapper in a bundled grouper tag, since red snapper is a more distinct fishery.

41. Telephone interview with Harry Upton (see note 11).

References

Arnason, Ragnar, and Peter Pearse. 2008. Allocation of Fishing Rights between Commercial and Recreational Fishers. This volume.

Coleman, Felicia C., Will F. Figueira, Jeffrey S. Ueland, and Larry B. Crowder. 2004. The Impact of United States Recreational Fisheries on Marine Fish Populations. *Science* 305(5692): 1958–60.

Committee on the Review of Recreational Fisheries Survey Methods (CRRFSM). 2006. *Review of Recreational Fisheries Survey Methods.* Washington, DC: National Academies Press.

Connecticut Department of Environmental Protection, Wildlife Division (CT DEP). 2005. *Connecticut Deer Program Summary: 2005.* Hartford: CDEP.

———. 2006. *2006 Hunting and Trapping Field Guide.* Hartford: CDEP.

Criddle, Keith. R. 2008. Examining the Interface between Commercial Fishing and Sportfishing: A Property Rights Perspective. This volume.

Criddle, Keith R., Mark Herrmann, S. Todd Lee, and Charles Hamel. 2003. Participation Decisions, Angler Welfare, and the Regional Economic Impact of Sportfishing. *Marine Resource Economics* 18(4): 291–312.

Florida Fish and Wildlife Conservation Commission (FWC). 2005. *2005-2006 Hunting Handbook.* Springville, UT: Liberty Press Publications. Downloaded May 2006 from http://wildflorida.org/hunting/pdf/05-06FL_Hunting_Handbook.pdf.

Freese, Curtis H., and David L. Trauger. 2000. Wildlife Markets and Biodiversity Conservation in North America. *Wildlife Society Bulletin* 28(1): 42–51.

Gislason, Gordon. 2006. Commercial vs Recreational Fisheries Allocation in Canada: Pacific Herring, Salmon and Halibut. Paper Presented at Sharing the Fish 2006 Conference, Perth, Western Australia, February 26–March 2.

Grafton, R. Quentin, Ragnar Arnason, Trond Bjørndal, David Campbell, Harry F. Campbell, Colin W. Clark, Robin Connor, Diane P. Dupont, Rögnvaldur Hannesson, Ray Hilborn, James E. Kirkley, Tom Kompas, Daniel E. Lane, Gordon R. Munro, Sean Pascoe, Dale Squires, Stein I. Steinshamn, Bruce R. Turris, and Quinn Weninger. 2006. Incentive-based Approaches to Sustainable Fisheries. *Canadian Journal of Fisheries and Aquatic Sciences* 63(3): 699–710.

Gulf of Mexico Fishery Management Council (GMFMC). 1984. *The Reef Fish Fishery Management Plan.* Tampa, FL: GMFMC.

———. 1997a. *Amendment Number 12 to the Reef Fish Fishery Management Plan.* Tampa, FL: GMFMC.

———. 1997b. *Amendment Number 14 to the Reef Fish Fishery Management Plan.* Tampa, FL: GMFMC.

———. 1999. *Amendment Number 16B to the Reef Fish Fishery Management Plan.* Tampa, FL: GMFMC.

———. 2003. *Corrected Amendment for a Charter Vessel/Headboat Permit Moratorium Amending the FMPs for: Reef Fish (Amendment 20) and Coastal Migratory Pelagics (Amendment 14) (Including EA/RIR/IRFA).* Tampa, FL: GMFMC.

———. 2005. *Recreational Fishing Regulations for Gulf of Mexico Federal Waters.* Tampa, FL: GMFMC. November.

Idaho Fish and Game (IFG). 2006. Idaho Big Game Season. Rules 2006. (Available at http://fishandgame.idaho.gov/cms/hunt/rules/bg.)

Johnston, Robert J., Daniel Holland, Vishwanie Maharaj, and Tammy W. Campson. 2008. Evaluation of Fish Tags as an Attenuated Rights-Based Management Approach for Gulf of Mexico Recreational Fisheries. *Connecticut Sea Grant Publication* CTSG-08-07. Groton, CT: Connecticut Sea Grant College Program.

Kim, Hwa Nyeon, Richard T. Woodward, and Wade L. Griffin. 2008. Can Transferable Rights Work in Recreational Fisheries? This volume.

Leal, Donald R., Michael De Alessi, and Pamela Baker. 2006. *Governing U.S. Fisheries with IFQs: A Guide for Federal Policymakers.* Bozeman, MT: PERC.

Leal, Donald R., and J. Bishop Grewell. 1999. *Hunting for Habitat: A Practical Guide to State-Landowner Partnerships.* Bozeman, MT: PERC.

Macinko, Seth, and Daniel W. Bromley. 2002. *Who Owns America's Fisheries?* Washington, DC: Island Press.

Mestl, Gerald E. 2001. 2000 Missouri River Ecology. *Performance Report* F-75-R-18. Lincoln, NE: Nebraska Game and Parks Commission, Fisheries Division.

Millard, Michael J., Stuart A. Welsh, John W. Fletcher, Jerre W. Mohler, Andrew W. Kahnle, and Kathryn A. Hattala. 2003. Mortality Associated with Catch and Release of Striped Bass in the Hudson River. *Fisheries Management and Ecology* 10: 295–300.

National Marine Fisheries Service (NMFS). 2006. *Gulf of Mexico Grouper Facts.* St. Petersburg, FL : U.S. Department of Commerce, National Oceanic and Atmospheric Administration, National Marine Fisheries Service, Southeast Regional Office.

——— (NMFS). 2007a. *Gulf of Mexico Grouper Management in Federal Waters: Frequently Asked Questions.* St. Petersburg, FL: U.S. Department of Commerce, National Oceanic and Atmospheric Administration, National Marine Fisheries Service, Southeast Regional Office.

——— (NMFS). 2007b. *Frequently Asked Questions: Scientific Review of Recent SEDAR Grouper Assessments.* St. Petersburg, FL: U.S. Department of Commerce, National Oceanic and Atmospheric Administration, National Marine Fisheries Service, Southeast Regional Office.

National Oceanic and Atmospheric Administration (NOAA). 2004. Fisheries of the Caribbean, Gulf of Mexico, and South Atlantic; Reef Fish Fishery of the Gulf of Mexico; Red Grouper Rebuilding Plan. *Federal Register* 69(114): 33315–21.

———. 2006. Fisheries of the Caribbean, Gulf of Mexico, and South Atlantic; Gulf of Mexico Recreational Grouper Fishery Management Measures. *Federal Register* 71(222): 66878–80.

National Research Council (NRC). Committee to Review Individual Fishing Quotas. 1999. *Sharing the Fish: Toward a National Policy on Individual Fishing Quotas.*

Washington, DC: National Academies Press.

North Pacific Fishery Management Council (NPFMC). 2006. *Charter Halibut Stakeholder Committee Recommendation for Permanent Solution Alternatives and Options.* Anchorage, AK: NPFMC. April.

Oregon Department of Fish and Wildlife (ODFW). 2005. *Oregon Big Game Regulations 2005.* (Available at http://www.dfw.state.or.us/resources/hunting/big_game/regulations/reg_book.pdf.)

Pennsylvania Game Commission (PGC). 2006. *Pennsylvania Hunting and Trapping Digest 2006 – 2007.* Harrisburg: PGC.

Schirripa, M. J., and Christopher M. Legault. 1999. Status of the Red Snapper in U.S. Waters of the Gulf of Mexico: Updated through 1998. *Contribution: SFD*-99/00-75. Miami, FL: U.S. Department of Commerce, National Oceanic and Atmospheric Administration, National Marine Fisheries Service, Southeast Fisheries Science Center, Sustainable Fisheries Division.

Scott, Anthony D. 1988. Conceptual Origins of Rights Based Fishing. In *Rights Based Fishing*, ed. Philip A. Neher, Ragnar Arnason, and Nina Mollet. Dordrecht, The Netherlands: Kluwer Academic Publishers, 11–38.

———. 2000. Introducing Property in Fishery Management. In *Use of Property Rights in Fisheries Management*, ed. Ross Shotton. Proceedings of the FishRights99 Conference. Fremantle, Western Australia, November 11–19, 1999. *FAO Fisheries Technical Paper* 404/1 and 404/2: 26-38. Rome: Food and Agriculture Organization of the United Nations, 1–13.

Smith, Phil, and Jane DiCosimo. 2006. *Chronology of NPFMC Actions to Manage the Charter Halibut Fishery.* Anchorage, AK: U.S. Department of Commerce, National Oceanic and Atmospheric Administration, National Marine Fisheries Service.

Stone, Clifton, and Jason S. Sorenson. 2002. *Paddlefish Use and Harvest Survey below Gavins Point Dam, South Dakota Utilizing a Limited Entry Tag/Permit System, 1997–2001: Progress Report.* Pierre: South Dakota Department of Game, Fish and Parks.

Sutinen, Jon G. 1993. Recreational and Commercial Fisheries Allocation with Costly Enforcement. *American Journal of Agricultural Economics* 75(5): 1183–87.

Sutinen, Jon G., and Robert J. Johnston. 2008. Angling Management Organizations: Integrating the Recreational Sector into Fishery Management. This volume.

Sutton, Stephen G., John R. Stoll, and Robert B. Ditton. 2001. Understanding Anglers' Willingness to Pay Increased Fishing License Fees. *Human Dimensions of Wildlife* 6: 115–30.

Taylor, Justin, and Thomas L. Marsh. 2003. Valuing Characteristics of Transferable Deer Hunting Permits in Kansas. Paper presented at the Western Agricultural Economics Association Annual Meeting, Denver, CO, July 11–15.

United States Fish and Wildlife Service (USFWS). 2002. *The Federal Duck Stamp Story.* Arlington, VA: USFWS, Federal Duck Stamp Office.

———. n.d. *The Federal Duck Stamp Program.* (Accessed 6/10/06 from http://www.fws.gov/duckstamps/.)

Woodward, Richard T., and Wade L. Griffin. 2003. Size and Bag Limits in Recreational Fisheries: Theoretical and Empirical Analysis. *Marine Resource Economics* 18(3): 239–62.

Chapter 9

Angling Management Organizations: Integrating the Recreational Sector into Fishery Management

Jon G. Sutinen and Robert J. Johnston

Recreational fishing is one of the more popular pastimes in the United States, and a significant amount of activity occurs in marine waters (USFWS 2002). In 2006, about thirteen million saltwater anglers made over eighty-nine million fishing trips to the Atlantic, Gulf, and Pacific coasts of the United States, and caught an estimated 476 million fish (NMFS 2007, 21). Marine recreational fishing activity (based on number of fishing trips per year) has been growing in most coastal areas of the United States (NMFS 2002b). It increased by over 20 percent from 1996 to 2000. Nearly a third of this growth occurred in the Gulf of Mexico region, followed closely by the Mid-Atlantic region (at just over a fourth), and the South Atlantic region (about one-fifth).

The dramatic rise in marine recreational fishing, however, has exacerbated conflicts with commercial fishers and depletion of fish stocks. The conflicts and resource depletion are threatening the future of recreational fisheries in the United States as well as in other countries. In this chapter we examine ways to rescue this future by reducing conflicts and improving the sustainability and value of marine recreational fisheries. We explore options for fully integrating the recreational sector into the management of fisheries. Among these is a novel approach called Angling Management Organizations (AMOs), which combines three of the more pervasive and promising trends in fishery management worldwide—management devolution, strengthened harvest rights, and

---

This chapter is a shortened version of Sutinen and Johnston (2003).

co-management. AMOs are community-based organizations loosely related to rights-based producer organizations in commercial fisheries. They are designed to strengthen resource stewardship, reduce enforcement and monitoring costs, alleviate management conflicts, and produce greater long-term net economic benefits in recreational fisheries. They conform to seven basic principles of integrated fishery management, which are described below. The other five organizational structures considered here, including the status quo, do not conform to all seven principles and are not expected to be as effective as AMOs.

Issues: Conflicts and Depletion

The expansion of fishing effort by both recreational and commercial sectors during the 1990s placed fish stocks under pressure in several U.S. fisheries (NMFS 2002a) and contributed to increasing conflict between the two sectors. According to the National Marine Fisheries Service (NMFS 2002a), the stocks of three of the ten most popular recreational marine species are not healthy. Overfishing is occurring and the stocks are overfished in the cases of red drum (South Atlantic and Gulf of Mexico) and scup (Mid-Atlantic). Overfishing is occurring for summer flounder (Mid-Atlantic), but the stock is not yet overfished. The bluefish stock in the Atlantic is overfished, but overfishing is not occurring.

Conflicts and disputes between the recreational and commercial fishing sectors often center on such factors as (1) the use of different management measures to manage the recreational and commercial sectors and (2) the explicit or de facto open-ended reallocation of harvest from the commercial sector to the recreational sector, or vice versa (Johnston and Sutinen 1999; NPFMC et al. 1998; NPFMC and ISER 1997).[1] In addition to conflicts with commercial fishers, the effects of a growing recreational sector have included (1) localized stock depletion in specific geographical areas favored by recreational anglers, (2) overcrowding of productive grounds and declining catches, (3) a potential race-to-fish in the recreational sector, and (4) conflicts between competing recreational user groups.

The sustainability and long-term social value of recreational fisheries are further threatened by ongoing trends in management. A recent report by the National Academy of Public Administration (NAPA) concludes that the federal fishery management system has "increasingly struggled under the burdens of conservation, environmental protection, over-exploitation, and increased statutory and policy mandates" (NAPA 2002, ix). Moreover, in the United States, the growth of litigation has diverted resources away from the basic tasks of fisheries management (Hogarth 2002; NAPA 2002).

The system that produces fishery management regulations is cumbersome and inflexible,[2] with a tendency to enact regulations that fishers view as overly complex and inappropriate for their fishery. For example, the federal council system tends to establish common rules for fishing activity over very large spatial scales. While this approach is partially justified on the basis of biological considerations, the use of the same broad spatial scale for establishing management rules threatens the ability of management to optimize socioeconomic objectives. In general, the broader the spatial scale, the more diverse the interests of the fishers and the greater the difficulty to design rules of fishing that are optimal for all, since acceptable compromises and consensus on common rules are difficult to achieve. These problems are not unique to the United States. Many countries have encountered similar difficulties and have concluded that heavy government involvement in fishery management is burdensome, inflexible, and ultimately ineffective (Pomeroy 1999).

Other notable trends with potential implications for recreational fisheries management include economic and demographic change, increased use of market mechanisms in fisheries, and the application of rights-based approaches to commercial fisheries management. These and other changes place increasing pressure on existing recreational management regimes (Sutinen, Johnston, and Shaw 2002).

An Example: The Gulf of Mexico Red Snapper Fishery

The Gulf of Mexico red snapper fishery serves as a good example of a mixed recreational–commercial fishery that is poorly served by existing management arrangements. It illustrates an archetype in which management has not adequately addressed three primary issues: (1) fishery overexploitation, (2) increasing commercial–recreational conflicts, and (3) heterogeneity in the recreational fishery. Red snapper is an important component of the large multi-species reef fish fishery in the Gulf of Mexico (GOM). It is one of the primary reef fish targeted by both commercial and recreational fishing sectors in the GOM. Red snapper also is subject to significant incidental catch by commercial shrimp trawls. As a result of combined directed and incidental harvest, GOM red snapper stocks have been placed under substantial and long-term fishing pressure. In the early 2000s, the stock was classified as both overfished and subject to overfishing (NMFS 2002a).

Red snapper harvests are currently far below their historical highs but have remained relatively stable over the previous decade. Recent years have witnessed red snapper mortality increasing in the recreational sector, remaining relatively constant in the commercial sector, and remaining

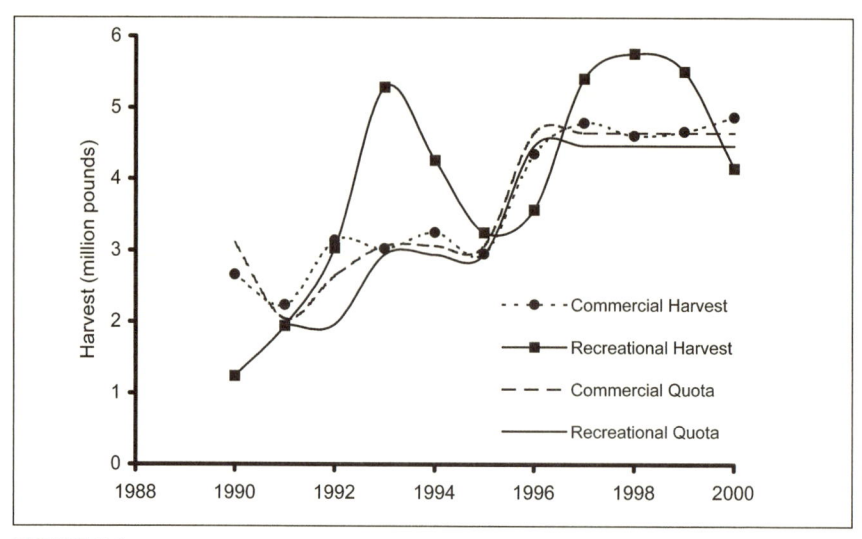

FIGURE 9.1
Gulf of Mexico Red Snapper Harvest and TAC (commercial and recreational sectors)
Source: NMFS (2006, 157–64).

constant in the shrimp bycatch sector (GMFMC 2001). Figure 9.1 illustrates commercial and recreational harvests from 1990–2000, along with the commercial and recreational total allowable catches (TACs) set by the Gulf of Mexico Fishery Management Council (GMFMC). While commercial harvests in general correspond with the TAC, recreational harvests have frequently exceeded the official TAC, often by significant margins. While the estimated stock of legal harvest-sized red snapper has increased in recent years, the total population shows no evidence of increase and may have diminished (GMFMC 2001).

Figure 9.2 illustrates the number of days in which the commercial and recreational red snapper fisheries were officially open. Both fishery sectors reveal patterns in which open days have diminished over time, even as the total harvest has remained relatively constant. Recent regulatory amendments reveal a trend toward smaller bag limits and increasing minimum length (Schirripa and Legault 1999). As in most marine recreational fisheries, the GMFMC sets management regulations over a wide spatial scale. For example, the recreational season for red snapper in the Gulf of Mexico, which runs from April 21 through October 31, applies throughout the Gulf, with no local or regional variation. The particular dates of the open season, however, may not provide optimal benefits to anglers in all geographic areas.

The magnitude of the GOM recreational red snapper fishery and performance of existing management indicates that significant economic gains may

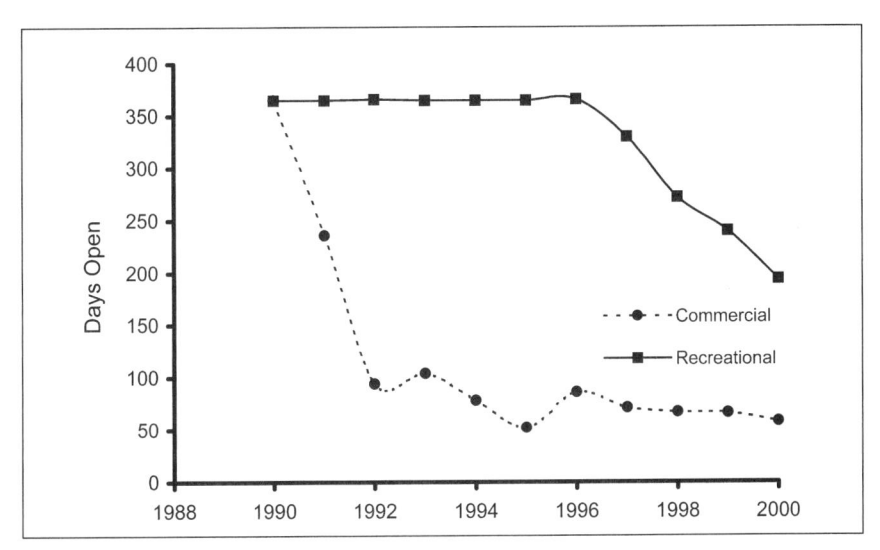

FIGURE 9.2
Gulf of Mexico Red Snapper Fishery Days Open (commercial and recreational sectors)
Source: NMFS (2006, 157–64).

be realized through management arrangements that successfully integrate the recreational sector into overall fishery management, control fishing mortality, and address the dispersion and heterogeneity characteristics of the recreational fishery.

Integrated Management

The recreational sector of a fishery is fully integrated into the fishery's management program when management measures applied to the recreational sector are sufficient to enable managers to achieve the goals of the fishery management plan (such as sustainability and socioeconomic objectives) and achieve the agreed upon allocation of catches among recreational, commercial, and other user groups. For example, the recreational sector would not be fully integrated into a fishery management program where the management measures provide little or only weak control over recreational fishing mortality or where the measures allow one sector to erode the amount of catch to which the other sector is entitled. This section develops a set of seven basic principles for improving the management of recreational fisheries (see figure 9.3). Each principle builds on the previous principle, and all are essential ingredients for fully realizing the benefits of integrated management.

Principle 1: Integrated recreational management is desirable only where the benefits of integration outweigh the costs of integration.

Principle 2: A workable mechanism must exist for allocating catches among recreational, commercial, and other user groups as a precondition for integrated recreational management.

Principle 3: Managers must implement management measures that in practice provide a high degree of control over recreational fishing mortality.

Principle 4: Recreational fishery management should be based on a system of strong angling rights.

Principle 5: Recreational fishery managers should consider assigning angling rights to organizations or other groups as well as to individuals in recreational fisheries.

Principle 6: Recreational fishery management should be decentralized, with limited management authority devolved to and shared with local organizations and governing institutions.

Principle 7: Cost recovery should be applied to recreational fishery management since it will strengthen accountability and improve the overall performance of the management program.

FIGURE 9.3
Seven Principles of Integrated Management

Despite the potential gains from integration, it may not be socially beneficial for managers to fully integrate the recreational sector into management for all fisheries. The primary element determining whether management should fully integrate the recreational sector is the balance between the social costs and benefits of integration. These, in turn, depend on the characteristics of the fishery. For example, if recreational catch accounts for a small proportion of total fishing mortality, full integration into the management program may not be warranted. Even where recreational fishing mortality is significant, the costs of full integration may make it undesirable on economic grounds. These costs include the costs of collecting and analyzing data on the sector as well as the costs of administering and enforcing the integrated program (Chapman, Blias, and Gooday 2001). This leads to Principle 1, that integrated recreational management is desirable only where the benefits of integration outweigh the costs of integration.

Managers must set a TAC consistent with sustainability of fishery resources, and allocate this TAC among user groups. The allocation of catch among user groups often is a highly contentious and controversial process. Integration of the recreational fishery into management does not eliminate issues related to initial TAC allocation. The mechanism for allocation—whether administrative or market-based—must have widespread support among stakeholders. This leads to Principle 2, that managers must develop a workable mechanism for allocating catches among recreational, commercial, and other user groups as a precondition for integrated recreational management.

Effective management of a fishery requires strong control of fishing mortality from all sources (commercial, recreational, subsistence, incidental, etc.). There are many fisheries in which managers exercise weak control of recreational catches, leading to conflict and unsustainable harvests. A hard or binding recreational TAC appears to offer the highest degree of control over recreational fishing mortality. This leads to Principle 3, that managers must implement management measures that in practice provide a high degree of control over recreational fishing mortality.

The current trends of decreasing open seasons and shrinking bag limits in recreational fisheries reflect the fundamental flaws in existing management measures. Such measures are ultimately doomed to failure because they cannot satisfactorily address pressures related to growth within the recreational sector itself and conflicts with other resource users. Under a system of strong angling rights, existing recreational anglers would be secure from the threat of new entrants into the fishery. The establishment of a secure harvest right would provide standing and precedent with which the interests of recreational fishery may be protected from those who might otherwise seek to appropriate fishery resources for their own use.[3]

Rights-based systems have a proven record of accomplishment in promoting sustainable management of fisheries and producing wealth. Rights-based systems effectively constrain exploitation within set limits, mitigate the race-to-fish, reduce overcapacity and gear conflicts, and improve product quality and availability. Producers benefit, consumers benefit, and, when the resource rent is used to pay for the cost of management, the general public benefits. In addition, there are environmental benefits that result from reduced fishing capacity (OECD 1997; Arnason 2001).

While recreational fisheries are in many ways different from their commercial counterparts, and the only significant evidence on rights-based fisheries experiences is from commercial fisheries, the sectors share many of the same concerns related to resource stewardship. As the evidence cited shows, stronger rights are superior to weaker rights.

This leads to Principle 4, that recreational fishery management should be based on a system of strong angling rights.

The distinguishing characteristics of recreational fisheries suggest that while rights-based approaches will likely offer many of the same advantages to the recreational sector as they have provided in commercial fisheries, the details of successful rights-based approaches may differ between the recreational and commercial sectors. For example, harvest rights need not be allocated to individual fishers. In response to controversies associated with individual fishing quotas (IFQs) in commercial fisheries,[4] some management authorities have developed group- and community-based alternatives that also feature strong harvest rights. Management authorities in Canada, Denmark, the Netherlands, Norway, Sweden, and the United Kingdom have allocated harvest rights to organizations of commercial fishers. Within the United States, Community Development Quotas and harvest cooperatives (both in Alaska) represent creative rights-based alternatives to IFQs.

These experiences lead us to propose Principle 5, that recreational fishery managers should consider assigning angling rights to organizations or other groups as well as to individuals in recreational fisheries.

National governments also are decentralizing fisheries management by devolving management authority to lower levels of government. In recent years, Canada, Denmark, the Netherlands, Norway, Sweden, the United Kingdom, and other countries have devolved fishing rights and duties to fishers and their organizations.[5] Japan has built on a lengthy tradition of rights-based management and now has the world's most extensive and sophisticated fisheries co-management system (NRC 1999). The trend toward devolved and shared management authority emphasizes local organizations and governing institutions (Pomeroy 1999).

Governments do not typically devolve all management authority, nor do they decentralize all management functions. The authority to make conservation decisions is nearly always kept in the hands of government and at a fairly centralized level. For example, the authority to establish TACs or to establish areas closed to fishing activity typically remains with the government. Socioeconomic decisions, on the other hand, are devolved to lower levels of government or to fishing organizations. For example, in the United States, the Atlantic States Marine Fisheries Commission, a governing body that oversees fisheries encompassing coastal waters of several states, sets TACs for several species and then allocates these TACs among coastal states. Individual coastal states choose various methods to remain within their individual TACs. In Europe, the European Commission sets TACs for numerous species, while individual coastal countries have the authority to manage their fishing fleets so as to maintain harvests within their TACs. In the UK, the government has further divided the country's TACs among producer organizations, and each producer organization is given the authority to manage its share of national TACs to optimize the socioeconomic objectives of its members. In Canada, the government has allocated TACs to community-based fishing organizations and authorized the local organizations to regulate its members to achieve collective socioeconomic objectives.[6]

Why do governments devolve management authority? These countries have found that the sharing of management authority with fishers—known as co-management—reduces administrative costs and greatly improves compliance with management regulations. Decentralized management has proved to be more effective and to produce more benefits than highly centralized management (Pomeroy 1999).

Both reasoning and evidence lead to Principle 6, that recreational fishery management can be improved by more decentralization, where limited management authority is devolved to and shared with local organizations and governing institutions. The key question is how to accomplish this in specific cases. We address this issue in depth below.

Who pays and how they pay for management services influences the performance of a fishery as well as the nature and extent of fisheries expenditures. Financing is often viewed as "merely" a distributional issue, but, in fact, sustainable financing has become an increasingly important issue not just to ensure that revenues cover costs but also as a way to affect incentives that encourage favorable behavior and discourage unfavorable actions. Cost-recovery measures have the potential to realize significant improvements in the overall performance of fisheries management. In other words, cost recovery can improve economic efficiency and conservation of fishery resources (Andersen and Sutinen 2003).

Efficiency gains can come from two sources. The first source of efficiency gains is the improved cost efficiency in the production of services, such as research, administration, and enforcement. The second source of efficiency gains is the production of a more valuable mix of management services—a mix that better reflects needs of the users. Simply put, government managers have less of an incentive to minimize costs associated with management and less of an ability to identify the mix of management services most valued by resource users. In the absence of well-defined fishing rights, government control over management services and research may be required, as individual users have little incentive to pursue activities or management methods that sustain the resource in question. However, within a strong rights-based system, users have a stronger stewardship incentive. In such cases, society typically gains when resource users both pay for and influence the set of management services or mechanisms applied.

We conclude this section with Principle 7, that cost recovery should be applied to recreational fishery management, since it will strengthen accountability and improve the overall performance of the management program.

Angling Management Organizations

How then can these principles be used to develop a rights-based system for recreational fisheries, given the challenges posed by recreational fisheries? As an illustration of how this can be done, we discuss a method we proposed for fully integrating the recreational sector into the management of red snapper in the Gulf of Mexico. Integrated recreational management appears to be highly desirable in this fishery. Recreational catches are significant in the fishery and are approximately comparable to commercial catches. Recreational catches have often exceeded the recreational TAC set by the GMFMC and, over time, the length of the open fishing season has diminished, bag limits have become smaller, and the minimum fish size has increased.

Principles 1, 2, and 3 are already satisfied in the fishery. Under the reef fish fishery management plan, a mechanism currently exists for allocating red snapper catches among recreational, commercial, and other user groups as a precondition for integrated recreational management. The plan sets a TAC for the recreational sector, which, in principle, provides a high degree of control over recreational fishing mortality.[7] The problem to date has been that the measures for implementing the TAC (bag limits and seasons) have not always succeeded in limiting recreational catch to the TAC limit (see Kim, Woodward, and Griffin, this volume).

At present, Principles 4–7 are not satisfied in the fishery. Principle 4 states that recreational fishery management should be based on a system of strong angling rights. Given the advantages of rights-based management and the potential costs and complications associated with individual recreational quotas (explained below), we propose establishment of a novel set of institutions that we shall entitle Angling Management Organizations (AMOs). AMO management conforms to the seven principles enumerated above and represents the combination of devolved co-management with rights-based fishery management. As detailed below, AMOs are appropriate for fisheries in which (1) the apparent benefits of integrated recreational management outweigh the apparent costs of integration, (2) a workable mechanism exists, or can be created, for allocating catches among all user groups, and (3) managers implement management measures that provide a high degree of control over fishing mortality—our first three principles of integrated management.

AMOs are nongovernmental organizations comprised of groups of recreational anglers (cf. Principle 5). Unlike traditional IFQ management, in which rights are assigned to individuals, here angling rights are assigned to AMOs through an assignment of a fixed share of the recreational TAC. Individual recreational anglers obtain the right to manage a proportion of the recreational harvest through ownership of shares in a particular AMO, much as one might own shares in a private company. After initial distribution, AMO shares may be bought and sold much like shares in companies are traded on a centralized stock market or exchange. Some of the more important attributes of the AMOs include: Each AMO has the exclusive right to determine how its share of the recreational TAC is used; it has the authority to implement measures to optimize socioeconomic objectives; it is a nongovernmental organization of anglers; it is financially independent and sustainable; and it provides equal opportunity to fish to all anglers.

The proposed system would devolve limited responsibility to AMOs—an application of co-management to the recreational fishery (cf. Principle 6). Each AMO would be jointly responsible for ensuring that its share of the TAC was not exceeded. The consequence of recreational TAC violation at the AMO level would be a reduction (either temporary or permanent) in its share of the recreational TAC. Hence, there would be strong incentives for self-policing of member anglers. Moreover, because each AMO—and by extension AMO shareholders—would have a strong right to a certain share of the total harvest, the advantages of rights-based management would be maintained.

In contrast, were standard IFQs to be applied to recreational fisheries (in which quotas were held by all individual anglers), management authorities would be responsible for overseeing the harvest of all anglers—a more costly form of centralized management. In addition, the highly dispersed and heterogeneous nature of recreational anglers would tend to complicate the use and enforcement of IFQ management. Unlike commercial fishers—typically fewer in number and with easily identifiable locations at which harvests are landed—recreational anglers are more numerous and may land harvests at locations that are more difficult to observe. Moreover, transaction costs associated with the management and trading of quota shares, as well as the integration of anglers into the quota management system, are likely to be greatly inflated as the number of individual recreational quota holders increases.[8]

AMOs satisfy all seven principles. AMOs represent a rights-based approach in which rights are held by well-defined angler groups. Moreover, both management authority and cost—to a limited extent—are devolved or decentralized to local organizations with a clear stake in the resource. The proposed AMOs can be viewed as logical extensions of many existing angling organizations and are not dissimilar from some producer organizations that now play active roles in the management of several commercial fisheries (such as the cooperatives described above). However, as explained below, there are some important differences in institutional structure and performance between the proposed AMOs and commercial producer organizations.

Rights and Duties of AMOs

Each AMO shall possess a bundle of rights and duties. The foremost right is the exclusive right to catch a share of the sustainable total catch of a species each year. Each AMO shall have the authority and duty to manage the organization's quota and would be responsible for developing and implementing appropriate controls on catch. AMOs would be authorized to administer and enforce management measures imposed on the fishery by the AMO and be bound to use all reasonable means to maintain the total annual harvest under its quota.

Ownership of AMO Shares

A group of shareholders would own and operate each AMO. AMOs shall be nongovernmental organizations that take the form of conventional business (for-profit) entities, such as a corporation, limited liability company, or limited partnership. Shares of the AMOs shall be publicly traded, primarily to insure that the rights of minority shareholders are adequately protected.

Individual shareholders do not directly own a quota or right to a certain quantity of harvest, however. There is a distinction between owning a share of an AMO and having the right to harvest a particular proportion of the quota under control of that AMO. In other words, individuals do not have exclusive authority over a share of the organization's quota. Rather, all AMO shareholders collectively share the authority to manage the quota, which may be distributed to individual anglers (both AMO members and nonmembers) through a variety of mechanisms. That is, the AMO as an organization has a strong right to a certain quantity of harvest. Ownership of the AMO, in turn, is held by shareholders.

Each AMO is free to determine the use of its quota that will provide the greatest profits to its shareholders, subject to certain rules that provide equal opportunity to all anglers. However, this opportunity, in turn, may be subject to the angler's willingness to pay a license, access, or other fee. For example, an AMO might (1) auction off the right to fish in certain areas, (2) sell fishing licenses or rights to a certain quantity of harvest, (3) conduct fee-based fishing tournaments, or (4) conduct lotteries for rights to harvest in prime fishing grounds, among other options. As long as the right to participate in these programs is open to all interested anglers, they would be allowable. The right to raise revenue from the fishery resource provides the critical strong incentive for resource stewardship that is the hallmark of the proposed AMO program.

This is directly analogous to private companies who manage capital resources for profit. For example, a company that owns a private fishing lake may charge individuals for the right to fish in that lake. In this case, equal opportunity implies that all individuals have an equal opportunity to purchase fishing rights for that lake; it does not imply that individuals may access the lake without charge or constraint. Similarly, private landowners may conduct lotteries for game harvests on their land. All hunters have an equal opportunity to participate in the lotteries, but only a limited number would be chosen for the hunt, and even then must pay an access fee.

With share ownership, stakeholders have a valuable stake or interest in the AMO and in the fishery. Share ownership provides the incentive for shareholders to utilize more efficiently the resource than they otherwise would under alternative institutional arrangements. Owners would face the full consequences of their decisions regarding management of the AMO. As a result, shareholders would have the incentive to manage efficiently the AMO and the fishery. Ownership of the AMO provides a strong incentive to find ways to maximize the value of recreational fisheries (management methods, access facilities, habitat, artificial reefs, etc.). As ownership may (initially) be distributed in some fashion to current fishery stakeholders, the right to benefit from the capital resource would be assigned to those currently using the resource. The subsequent ability to buy and sell AMO shares will provide the ability for others to "buy in" to any AMO.

Under AMOs, recreational fish stocks would be viewed as capital assets. The owners of the rights to use the assets (AMO shareholders) will seek to maximize the value of recreational fishing in order to maximize the value of their portfolio of these assets. Ownership of AMO shares is important also for creating incentives to develop and use meaningful sources of AMO revenue. Otherwise, the pressure for creative solutions is not present and members would constantly be approaching government for subsidies and other forms of support.

Share ownership is superior to other forms of ownership. If the AMO were to hold the quota in trust for anglers, there would be no strong incentive to invest in the fishery and improve the benefits of fishing. A trust would not necessarily attract the most able managers and decision makers. Hard decisions would be postponed or avoided and significant achievements rare. Rent seeking would sap resources and the value of recreational fishing would suffer.[9] The same problems would occur if the government, whether local, regional or national, were the trustee. Hence, private ownership of shares is the superior alternative.

Membership in AMOs

AMO membership shall consist of voting members and nonvoting members. The voting members shall be the shareholders of the AMO. Nonvoting members could attend meetings and serve on AMO committees at the discretion of voting members. AMOs are expected to encourage participation by other stakeholders in order to resolve conflicts or identify and avoid conflicts before they arise. Voting is proportional to the number of shares owned. The AMO can decide to issue shares or to reduce the minimum number of shares held in order to increase its membership. Membership ought to be open, with all citizens eligible to acquire shares.

Management Measures and Authority

The AMO shall have the authority to manage the organization's quota. With that authority, the AMO will be tempted to use input controls and management measures other than the quantity of fish taken. To what extent should this authority be granted? A management measure, such as a ban on a certain fishing method, would be considered legitimate and acceptable if it were done for sustainability purposes. However, if the measure is used solely for allocation (i.e., to exclude or disadvantage some users) it is of questionable merit and must be discouraged or banned.[10]

Access to the Fishery under AMO Management

All anglers ought to have equal opportunity to acquire a unit right of access to the fishery. However, the AMO must have the ability to restrict access to the fishery in order to manage its quota effectively. The right to fish (defined in terms of fish caught, days fished, etc.) can be allocated to a limited number of anglers so long as each interested angler has a fair and equal opportunity to acquire the right. It is critical that AMO guidelines are established to prevent intentional or unintentional disenfranchisement of certain angler groups. While AMOs would have the right to charge fees for fishery access, it is important that these fees do not become a de facto barrier to entry, preventing fishery access to less wealthy anglers. Hence, the structure of allowable fees, along with any limitations on those fees to encourage equitable access, would be specified prior to the establishment of any AMO.

Characteristics of Quota and Shares

The unit of quota—whether in weight or numbers of fish—is one of the more important issues. It would likely be best to specify recreational quota in terms of numbers of fish rather than in terms of weight.[11] Basing quota on numbers of fish greatly simplifies offloading, reporting, and enforcement by eliminating the weighing requirement. In addition, most recreational anglers are familiar and comfortable with management measures based on numbers, rather than weight, of fish landed. Limits on pounds of fish landed are not commonly used in recreational fisheries because of the higher number of vessel landings and dispersed nature of the fishery. Moreover, since sport-caught fish are not bought or sold, it is impractical and expensive to have enforceable weigh stations at all sites of sport landings (NPFMC 2001). For these reasons, it is anticipated that quotas should be denominated in numbers of fish rather than by weight.

Other important characteristics of quotas and shares include the duration, divisibility, and transferability of both quota and AMO shares. Greatest security and value of rights are provided by those with a certain, indefinite duration, complete divisibility, and transferability. In most instances, there is no reason to specify these characteristics to be different from those required for commercial quota or for shares in any legal company.[12]

Quota Trading

Rights are stronger when no or few restrictions are placed on quota trading. Trades of any quantity of quota ought to be allowed at any time and between any quota holders. In other words, quota trades should be allowed among AMOs as well as between AMOs and commercial fishers. However, while unrestricted trading of quota among different AMOs would offer the potential for the greatest net benefits of the fishery resource, it could also risk the concentration of harvest shares in a particular geographic region, leading to potential localized depletion. Moreover, anglers and managers are likely to be at first unfamiliar and uncomfortable with unconstrained trading. Accordingly, some restrictions on quota trading may be desirable. As an example, the plan for charter IFQs in the Alaska halibut fishery allows charter boat operators to purchase IFQ shares from the commercial fishery, but shares originally allocated to the charter sector cannot be sold to the commercial sector. Constraints such as this are sometimes required to insure acceptance and/or a smooth transition to the implementation of a quota trading scheme and to prevent localized stock depletion. After a trial period (e.g., five years) restrictions ought to be evaluated to determine whether they should be relaxed to further strengthen angling rights.

Funding and Sustainability of AMOs

AMOs have the potential to be financially independent. Indeed, fiscal independence and sustainability are critical features of the AMO structure. AMOs must be fully responsible for raising their own revenue and covering their own costs. Otherwise, inefficiencies are certain to occur. The need to be fiscally sustainable creates a potent incentive to maximize the value of the fishery net of the costs of management.

To conform with Principle 7, each AMO shall have the right to raise revenue to cover the cost of its management program (research, enforcement, and administration) in several ways. One is to lease quota to parties who are willing to pay for exceptional rights to the quota, such as charter boats, fishing tournament organizers, and commercial fishers. Another is to authorize the AMO to impose levies on users of the resource (license, fuel, tackle, etc.). The AMO could impose a surcharge on share trades to raise revenue. Additional shares can be sold to raise funds. The AMO would be given exclusive authority to organize fishing competitions as well as to receive and sell the fish caught in said competitions if it so chooses.

Each AMO ought to have the authority to decide whether to allow anglers to sell their catch. Allowing sale of fish results in the more efficient utilization of the resource. An angler who catches some fish and who values the fish less than others in society is improving social welfare by selling the fish. Being able to sell fish also enhances the value of the recreational fishery. An AMO may choose to allow sale of recreational fish only at specific outlets and collect a levy on the sale. However, allowing sale encourages anglers to change the recreational nature of the fishery. If no comprehensive reporting system is in place, selling also may tend to weaken compliance and create serious enforcement problems.

Spatial Attributes and Considerations

The quota rights held by each AMO ought to be defined for a specific spatial area as well as the quantity of each species that may be caught. This would enable each AMO to limit the number of anglers and other users who have access to the area. Having the right to control access to an area will allow AMOs to alleviate conflicts among user groups and fishing methods and to avoid depletion of favorite fishing locations. Specifying rights in terms of an area and quantity of fish strengthens the exclusivity of the angling right, which will lead to more efficient utilization of marine resources and less conflict among users of the resources. This would also insure that efforts to displace anglers from operating in some areas could be done only with compensation.

There would be a variety of options for addressing spatial characteristics of AMOs. For example, AMOs might be defined purely geographically, such that only one AMO could operate in a specific geographic area. In contrast, one might allow multiple AMOs in a single region, defined by harvest category (e.g., charter boat vs. individual angler). The process must also address the issue of what species to assign to each AMO. Note that any spatial exclusivity of AMOs applies only to recreational fishing. It would be possible, indeed likely, that commercial fishing activity would overlap with spatial areas for which exclusive recreational fishing rights were held by an AMO.

Implementation Issues

AMOs represent a fundamental change in recreational fishery management. While the use of rights-based management and co-management in commercial fisheries is becoming more common, they have not yet been applied to recreational fisheries in a comprehensive manner. The development of the proposed AMOs would create a paradigmatic shift in the approach to recreational management—unlike the management changes typical in recreational fisheries, which often involve marginal adjustments in existing input or output controls.

Implementation of AMOs raises a number of issues, including eligibility criteria and means of selecting share recipients, disposing of AMO shares and quotas, characteristics of the shares and quota (extent and duration, divisibility, transferability), distribution of AMO shares and quotas among recipients, the size and scope of an AMO, monitoring and enforcement, and how to facilitate the transition from the status quo to the proposed AMO structure. These issues must be addressed in the process of designing and implementing the AMO organizational structure.

Initial Allocation and Disposition of Quota Rights; Selecting Eligible Recipients[13]

Once the recreational TAC is set for each recreational species, provisional AMOs would be initially formed for each recreational TAC. For example, a provisional AMO would be formed for each community-based or port-based share of the recreational TACs, say, for red snapper and other recreationally desired reef fish species in an area. Each AMO would issue shares and allocate the shares to eligible recipients. Shareholders collectively own the AMO and its quota holdings.

Defining the eligibility criteria for the initial allocation of AMO shares is a critical implementation step. The eligibility criteria will determine the set of AMO owners who, as a group, would receive the rights to the organization's recreational fishing quota. The principal consideration when setting eligibility criteria ought to be that they are perceived as fair by anglers and the larger public. Eligibility criteria for the initial allocation of AMO shares not perceived as fair would likely result in great resistance to the institution and might defeat it altogether.

So long as the shares can be traded and the cost of share trading is not great, the eligibility criteria for initial allocation are not expected to affect the long-term performance of AMOs. After a period of initial trading, the individuals who have the greatest to gain from share ownership will acquire the shares and will manage the AMO to produce maximum net value from the fishery.

At least two options are evident for determination of share eligibility. One is to make all citizens eligible, regardless of whether they are anglers. Another option is to restrict eligibility to those who can demonstrate past participation in a marine recreational fishery, such as by evidence of having held a saltwater license or of having owned recreational fishing gear and equipment, or other relevant evidence. Obviously, establishing a set of eligible recipients under the latter option could be time consuming and costly.

Initial Allocation of AMO Shares

Once the eligibility criteria for the initial allocation of AMO shares are set, the next step is to allocate AMO shares among eligible recipients. This requires decisions to be made regarding the number of initial shares in each AMO and the selection of initial recipients from the eligible pool of people. Both fairness and efficiency must be considered in these decisions. Anglers in general must perceive the method of initial allocation as fair. Otherwise the legitimacy of the AMO system would suffer. A legitimate system is needed to support voluntary compliance and to protect the system from political meddling. The initial allocation method also must insure that the AMO can be managed and operated cost effectively. A very large number of small shareholders may not have the incentive to run the AMO effectively. Moreover, voting and share trades may be costly and problematic under such circumstances.

One initial allocation option is to set the number of shares equal to the number of anglers in the fisheries over which the AMO has jurisdiction. Another option would be to allow anglers to nominate themselves to receive shares in one or more AMOs of their choice. Many casual anglers probably would not bother to nominate themselves, thus reducing the number of shareholders in each AMO. A third option would be to fix the number of shares in each AMO and distribute them to anglers using a lottery. Eligible anglers would place their name in a drawing for the AMO(s) of their choice. Names of new shareholders would be drawn at random for each AMO.

Scope and Size of Angling Management Organizations

The optimal scope and size is likely to be different for each AMO. Also, there is no reliable way to specify the optimal scope and size ex ante. AMO members will have the collective interest to maximize the value of the recreational fisheries under its jurisdiction. Over time the inherent economies of scope, size, and spatial area will operate to induce AMOs to merge or split into optimal entities. For example, the members of an AMO that is too large may find it in their interest to divide into two or more separate AMOs in order to better optimize their individual socioeconomic objectives and economize on costs of administration, research, monitoring, etc. Therefore, the law and regulations should facilitate AMOs changing their scope and size, so that each AMO finds its optimum in the course of time.

There is clear risk in allowing a large number of interest groups to establish separate AMOs—the result of such fractionalization could be AMOs insufficient to sustain required trading and regulatory activity. Hence, as a general guide, we recommend that AMOs be designed initially to be too large rather than too small. Rules should be established to minimize the costs of splitting into smaller AMOs where such division would increase net benefits to participating members.

Relationship to Existing Management Structures

To be effective, it will be necessary for AMO management to supplant existing management and regulatory structures. The replacement of existing management structures will likely require phase-in of some duration. For example, existing for-hire recreational vessels have access to the red snapper fishery by permit. Clearly, retaining the validity of these permits would reduce the incentive to join AMOs and would markedly reduce the efficacy of AMO management. Hence, just as in cases in which other types of rights-based management systems have been established worldwide, the proposed AMO system would replace existing permits and rights in the fishery. Existing recreational fishing permits and licenses would become invalid as AMO management was established.

Monitoring and Enforcement

The proposed AMOs are not expected to pose enforcement burdens and threats to compliance that are significantly greater than alternative management regimes. As indicated above, each AMO would have the authority and obligation to manage its share of the recreational TAC as well as to administer and enforce management measures to maintain the total annual harvest under the organization's quota. AMOs' advantage over alternative regimes is that shareholders have the incentive to protect their capital assets by monitoring the fishery and enforcing management measures. As anglers, shareholders themselves will tend to be more compliant and also tend to apply social pressure on others to comply. To maximize voluntary compliance, AMOs will likely find it in their interest to adopt procedures for developing and implementing policy that are open and perceived by the fishing public to be fair (cf. Sutinen 1996a).

Transition Period and Transaction Costs

Making the transition from Gulf-wide management by the GMFMC to shared management with a network of AMOs will present many issues that need to be addressed. AMOs will have to be organized and chartered. Each AMO would have to develop a means to develop and implement management policy. As in any management transition, the transition from current management to AMO-based management would involve substantial transaction costs. Accordingly, one of the primary considerations in considering whether AMOs are desirable in a given fishery is the magnitude of these costs relative to AMO benefits (see Principle 1 above). The same considerations apply to more traditional fishery co-management changes (Viswanathan, Abdullah, and Pomeroy 1998).

The approach to implementation that we propose, however, is not dependent on a prior estimate of costs and benefits. We propose an approach that, in its early stages, is not costly to implement, and relies on the beneficiaries of integration to take the initial steps toward integration. In other words, the parties who can best assess the relative costs and benefits of integration are provided the authority to determine the extent and nature of the integration measures that are ultimately implemented.

In addition to a process that minimizes initial transaction costs, it is critical that the process of establishing any new management structure be widely viewed as transparent, open, and fair, while not resulting in transaction costs that outweigh the potential benefits of management change (Viswanathan, Abdullah, and Pomeroy 1998). AMO implementation ought to encourage maximum stakeholder participation to resist "capture" of the process by any single stakeholder group or entrenched interest. However, it is equally critical that the process not be paralyzed by the contentions of too many narrowly defined interests. Elsewhere (Johnston and Sutinen 2003), we outline a detailed strategy for integration and transition, based on prior works that address the implementation of fisheries co-management worldwide.

We propose establishing a transition authority (TA) to explore and facilitate the formation of provisional AMOs. The TA would be appointed by and directly responsible to the Secretary of Commerce. The secretary would most likely delegate to an NOAA official in the region the authority to represent the department on this matter, and the TA would have a small number of members with appropriate expertise as well as a support staff and budget. The

TA would be given a specified amount of time (e.g., two years) in which to complete the task of forming the regional network of AMOs, in coordination with preexisting angler organizations and representatives. The first step in this process would be the establishment of a guided dialog among recreational and commercial fishers, other stakeholders, and regulators—a relatively low-cost process that would assess the potential role of AMOs in the fishery. From this initial step, further stages of implementation would proceed as outlined elsewhere (Johnston and Sutinen 2003).

The members of the TA should include at least one person with privatization experience, one with experience managing a private business, one with strong recreational interest, and one with in-depth knowledge of the fishery management system. The authority would receive a special funding appropriation to carry out its terms of reference.

Expected Achievements

The set of institutional arrangements proposed here for recreational fishery management are conceptually sound and provide an option for full integration of the recreational and commercial fishing sectors in fishery management. In addition, the incorporation of strong fishing rights associated with AMOs offers to encourage sustainable, efficient, and financially sound utilization of fishery resources. Devolution of both management authority and expense provides incentives for superior and cost-effective management, greater levels of voluntary compliance, and the ability to tailor management mechanisms to the socioeconomic goals of specific regions. Finally, paired with rights-based approaches in the commercial sector, AMOs would provide both a greater balance of influence and power among stakeholders in political and commercial marketplaces and a superior means of resolving conflicts among stakeholders.

Members of AMOs have a direct stake in the outcomes of their policies since they are owners of the community-based organization. The value of each AMO's shares directly reflects the performance of its policies and the value that anglers place on its approach to managing recreational TACs. That is, recreational fishery managers must face the consequences of their decisions, which align their private interests with the public interests.

Alternatives to AMOs

We examine here five alternatives to AMOs, including the status quo. As discussed, each of the five alternatives has some advantages but is inferior to AMOs in terms of satisfying all the seven principles of integration developed above.

Sub-Regional Recreational Management Councils

The first alternative to AMOs is to establish sub-regional recreational management councils. The sub-regional recreational fisheries management councils could be either modeled after or operate as subunits of the regional (parent) management councils. The sub-regional councils would operate under the same legal authority as the parent council, except that the sub-regional councils would not be charged with conservation decisions, which would be retained by the regional council. After setting TACs for recreational species, the parent council would allocate the recreational TACs among the sub-regional councils. The principal task of each sub-regional council would be to manage their share of the recreational TAC by setting management measures to optimize the region's socioeconomic objectives.

A closely related variant of this option would be the allocation of shares of regional recreational TACs among individual states (rather than sub-regional councils), and assigning to each state the authority to manage their share of the recreational TAC.[14] Each state would have the option to further allocate recreational TAC shares to local communities and delegate associated management authority to appropriate local government bodies. Each state or local government would be free to establish its own regulations, provided that the state's recreational catch does not exceed its TAC.

There are some distinct advantages to this decentralized structure. First, it separates the conservation decision from ongoing allocation decisions (save for the initial allocation of TACs among the sub-regional councils or states). This separation has been recommended for the federal fishery management system by several observers (e.g., Sissenwine and Mace 2001; Sissenwine and Rosenberg 1993), and is expected to encourage a more precautionary approach to conservation.

The principal disadvantages are related to Principles 4–7 above, which are not satisfied by this option. This option does not strengthen angling rights and, as a result, is not expected to improve the anglers' motivation to conserve stocks and enhance the value of recreational fisheries. While this option does have the merit of some degree of decentralization, management authority is not shared with local institutions.

In addition, this option has no provisions for strengthening accountability among managers and anglers. Sub-regional council members (or managers at the state level) have no direct stake in the outcomes of their policies. Councils or state management authorities neither reap the benefits of adding value to their recreational fisheries nor bear the costs of their proposed management actions, and there are no provisions for financial independence and fiscal sustainability.

Sub-Regional Advisory Committees

This option involves the establishment of a set of sub-regional advisory committees to recommend recreational fisheries policies to regional councils. A system of sub-regional advisory committees would function somewhat similarly to the sub-regional councils described above. The principal difference is that the sub-regional committees would operate within the structure of the regional councils in much the same way as do current advisory committees.

As with the previous option, the regional council would allocate the recreational TACs among the sub-regions. The principal task of each sub-regional advisory committee would be to recommend to the council a set of measures to optimize the region's socioeconomic objectives, while maintaining catches within the sub-regional TAC.

This option has all the disadvantages of sub-regional councils, as noted above. Moreover, unlike sub-regional councils, the sub-regional advisory option lacks the advantage of separating conservation decisions from allocation decisions. Hence, this option appears inferior to either AMOs or sub-regional management councils. However, the increased flexibility afforded by sub-regional advisement may increase net benefits compared to the status quo.

Individual Fishing Quotas

If feasible, individual recreational fishing quotas would achieve complete integration of the recreational and commercial sectors in the fishery management system. There are many reasons to favor such a management system, among them the seamless integration of recreational and commercial harvests and the elimination of the need to revisit commercial–recreational harvest allocation decisions.

Despite the potential advantages of recreational fishing quotas, the establishment and implementation of fishing quotas for individual recreational anglers face at least two formidable problems that likely render this option infeasible. The first problem, though not insurmountable in theory, is the initial allocation of quota among anglers. Since catch histories are nonexistent for most, if not all, recreational anglers, the most common basis for initial quota allocation cannot be used. While other means of initial allocation may be acceptable, this remains a fundamental challenge.

The second, and perhaps more formidable problem, is enforcement. Detecting noncompliance among recreational anglers is typically more difficult and costly than for their commercial counterparts (Sutinen 1993). Individual recreational quotas can only aggravate these problems since thousands of individuals' catches would have to be monitored. Enforcement, in all likelihood, would be ineffective. The regime would have to rely heavily on voluntary compliance. Extensive voluntary compliance with individual quotas could arise only with widespread support for IFQs among anglers, an unlikely prospect in the near term.[15]

IFQs for the for-hire sector (charter, party, and guide boats) of certain recreational fisheries, however, may be a feasible and desirable option and could supplement AMOs applied to individual anglers. Problems of initial allocation and enforcement, while still significant, are not expected to be more severe in the for-hire sector than in the commercial sector. The Alaskan halibut charter IFQ program is an example of such a program (NPFMC et al. 1998; Criddle, this volume).[16] This reflects the important distinction between management measures most appropriate for the for-hire sector and those most appropriate for individual anglers.

Status Quo

The status quo, in which full management authority rests with the GMFMC and the federal government, is clearly inferior to the AMO alternative described in this paper. The status quo fails to satisfy Principles 4–7, and shares all of the disadvantages of the first three options (sub-regional councils, state councils, and sub-regional advisory committees) described above. In addition, as a Gulf-wide, centralized management authority, the GMFMC cannot easily decide upon and implement customized rules across the varied interests represented in the coastal communities of the Gulf of Mexico.

Conclusion

A novel, practical option for recreational fishery management is the Angling Management Organization or AMO. Unlike other existing and proposed management alternatives, AMOs satisfy each of the seven principles of integration. As a result, AMOs are expected to encourage improved resource stewardship, reduced enforcement and monitoring costs, fewer management conflicts, and greater long-term net economic benefits in recreational fisheries.

AMOs are logical extensions of many existing recreational angling organizations and are not dissimilar from some of the organizations of producers that now play active roles in the management of several commercial fisheries. However, there are some important differences in institutional structure and performance between the proposed AMOs and commercial producer organizations. Some of the more primary attributes of AMOs are that each has the exclusive right to determine how to use its share of the recreational TAC; it has the authority to implement measures to optimize socioeconomic objectives; it is a nongovernmental organization of anglers; it is financially independent and sustainable; and it provides equal opportunity to fish to all anglers.

AMOs represent a paradigmatic shift in recreational fishery management, involving significant initial transaction costs. These costs notwithstanding, the desirable mix of incentives created by the proposed structure of these community-based organizations should provide a mechanism to decrease management costs in the long run, while markedly increasing recreational fishery benefits.

Notes

1. For example, the commercial sector may be subject to a hard or binding TAC, in which the commercial fishery is closed when the quota is met, and any overages are subtracted from future seasons' commercial harvest. In contrast, the recreational sector may be subject to a "target" TAC, in which violation results in neither closure of the fishery nor explicit deduction from the following year's recreational TAC. Rather, overages in the recreational sector are deducted from the next season's *total* TAC. This management practice imposes an indirect penalty on the commercial sector because the commercial sector receives a percentage of the total TAC, which is reduced by the overages by the recreational sector.

2. According to Kammer (2002), the National Marine Fisheries Service is the fourth largest regulatory regime in government—behind the Environmental Protection Agency, the Federal Aviation Administration, and the Federal Communications Commission. There is no uniformity in creation of regulatory records across the eight NMFS regions; and each regulatory decision endures eleven levels of review within NOAA.

3. For example, commercial fishers may view a growing recreational sector as a threat and may seek to influence policy so as to limit total recreational harvest. In fisheries with commercial IFQs, owners may argue that they have more at stake in a fishery than anglers without such rights. In such cases, the commercial sector tends to have a stronger voice, and the recreational sector a weaker voice, in matters of research and management policy (Sutinen 1996b; Pearse 1991). Currently, the recreational fishery has little means—outside of costly lobbying activities—to protect itself from these threats and to ensure the long-term sustainability of the fishery. The establishment of a rights-based management system would provide a clear incentive for stewardship within the recreational sector and a stronger legal basis for protecting the interests of those who benefit through recreational fishing activities.

4. Such controversies often surround the initial allocation of quota among individual producers, which can create class divisions in fishing communities, potential

threats to the way of life in coastal communities, and the claim by some that the government should not give away publicly owned fish resources.

5. We use the term devolution to refer to the act or process of shifting some decision-making and implementation authority to lower levels of government and to nongovernmental organizations and individuals. Managing fisheries requires that management authority is distributed across several levels of government and users of the resource. Co-management is where this authority is shared with users of the resource (Berkes et al. 2001).

6. See, for example, James (2002) for an excellent analysis of the experience with co-management in the fisheries of British Columbia, Canada.

7. This does not explicitly address the issue of recreational high-grading, discard mortality, or the potential mortality associated with catch and release fishing. Schirripa (1998) and Schirripa and Legault (1997 and 1999) illustrate that in recent years, a substantial portion of the total red snapper harvests are subsequently released. However, problems of discard mortality under the proposed AMO management are expected to be no greater than exist under current management. Moreover, the increased stewardship incentives (among recreational anglers) encouraged by the rights-based attributes of AMOs may diminish discard mortality.

8. As explained below, IFQs may be appropriate for the for-hire sector of a recreational fishery.

9. Rent seeking is a term to describe actions by individuals and interest groups designed to restructure public policy in a manner that will either directly or indirectly redistribute more benefits to themselves (Buchanan, Tollison, and Tullock 1980).

10. In addition, the usual provisions of business laws in most states would protect the rights of minority shareholders in such circumstances.

11. The plan for charter IFQs in the Alaska halibut fishery involves issuing quota in numbers of fish. The Alaska Department of Fish and Game will translate numbers to pounds based on an average weight estimate.

12. Some of the other characteristics of quotas and shares are discussed above.

13. For examples of how managers have addressed the challenges of initial allocation of individual quota in commercial fisheries, see Shotton (2001).

14. This is similar to the way in which the Gulf States Marine Fisheries Commission manages interstate fisheries. See, for example, *The Spotted Seatrout Regional Fishery Management Plan* (Blanchet et al. 2001).

15. Even if the problems of initial allocation and enforcement were readily solved, the added benefits of individual quotas compared to the proposed group quotas may not be significant. In commercial fisheries, individual quotas effectively mitigate the race-to-fish, a damaging feature of competitive TACs and other non-rights-based management measures. The race-to-fish in commercial fisheries significantly increases costs of production, post-harvest losses, bycatch and discards, and worsens product quality and safety. These same problems are not expected to be significant in recreational fisheries subject to community-based quotas.

16. It is important to note that the Alaska IFQ program only applies to charter anglers—non-charter anglers are not incorporated into the IFQ program.

References

Andersen, Peder, and Jon G. Sutinen. 2003. Financing Fishery Management: Principles and Economic Implications. In *The Cost of Fisheries Management*, ed. William E. Schrank, Ragnar Arnason, and Rögnvaldur Hannesson. Burlington, VT: Ashgate Publishing, 45–65.

Arnason, Ragnar. 2001. *Review of Experiences with ITQs: A Report for CEMARE*. Reykjavik: Department of Economics, University of Iceland.

Berkes, Fikret, Robin Mahon, Patrick McConney, Richard Pollnac, and Robert Pomeroy. 2001. *Managing Small-scale Fisheries: Alternative Directions and Methods*. Ottawa, ON, Canada: International Development Research Center.

Blanchet, Harry, Mark Van Hoose, Larry McEachron, Bob Muller, James Warren, Joe Gill, Terry Waldrop, Jerald Waller, Charles Adams, Robert B. Ditton, Dale Shively, and Steve VanderKooy. 2001. *The Spotted Seatrout Fishery of the Gulf of Mexico, United States: A Regional Management Plan*. Ocean Springs, MS: Gulf States Marine Fisheries Commission.

Buchanan, James M., Robert D. Tollison, and Gordon Tullock, eds. 1980. *Toward a Theory of the Rent-Seeking Society*. College Station: Texas A&M University Press.

Chapman, Lisa, Athena Blias, and Peter Gooday. 2001. *Incorporating the Recreational Sector in the Management of Australia's Commonwealth Fisheries: ABARE Report to the Fisheries Resources Research Fund*. Canberra: Australian Bureau of Agriculture and Resources Economics, October.

Criddle, Keith R. 2008. Examining the Interface between Commercial Fishing and Sportfishing: A Property Rights Perspective. This volume.

Gulf of Mexico Fishery Management Council (GMFMC). 2001. *Regulatory Amendment to the Reef Fish Fishery Management Plan to Set a Red Snapper Rebuilding Plan through 2032*. Tampa, FL: GMFMC.

Hogarth, William T. 2002. Testimony of Dr. William T. Hogarth, National Oceanic and Atmospheric Administration, Assistant Administrator for Fisheries, before the President's Commission on Ocean Policy, Charleston, SC, January 15. (Accessed 12/6/07 from http://www.nmfs.noaa.gov/OceanCommHogarthTestimony011502.htm.)

James, Michelle. 2002. Co-management and Beyond: The British Columbia Experience. Paper presented at the International Institute of Fisheries Economics and Trade 2002 Conference, Wellington, New Zealand, August 19–22.

Johnston, Robert J., and Jon G. Sutinen. 1999. *Appropriate and Inappropriate Economic Analysis for Allocation Decisions: The Case of Alaska Halibut*. Research paper prepared for the Halibut Coalition, Juneau, AK.

Johnston, Robert J., and Jon G. Sutinen. 2003. *Angling Management Organizations: An Option for Fully Integrating the Recreational Sector into the Management Program for the Gulf of Mexico Red Snapper Fishery*. Research paper prepared for Environmental Defense, Austin, TX.

Kammer, Ray. 2002. Testimony before the U.S. Senate Committee on Commerce, Science and Transportation, Subcommittee on Oceans, Atmosphere and Fisheries. May 8.

Kim, Hwa Nyeon, Richard T. Woodward, and Wade L. Griffin. 2008. Can Transferable Rights Work in Recreational Fisheries? This volume.

National Academy of Public Administration (NAPA). 2002. *Courts, Congress, and Constituencies: Managing Fisheries by Default.* A Report by a Panel of the National Academy of Public Administration for the Congress and the U.S. Department of Commerce, National Marine Fisheries Service. Washington, DC: NAPA.

National Marine Fisheries Service (NMFS). 2002a. *Toward Rebuilding America's Marine Fisheries: Annual Report to Congress on the Status of U.S. Fisheries—2001.* Silver Spring, MD: U.S. Department of Commerce, NOAA, National Marine Fisheries Service, Office of Sustainable Fisheries.

———. 2002b. *Recreational Fishing Statistics Queries.* (Available at http://www.st.nmfs.gov/st1/recreational/queries/index.html.)

———. 2006. *Final Supplemental Environmental Impact Statement for Amendment 26 to the Gulf Of Mexico Reef Fish Fishery Management Plan to Establish a Red Snapper Individual Fishing Quota Program (Including a Revised Initial Regulatory Flexibility Analysis and Regulatory Impact Review).* St. Petersburg, FL: U.S. Department of Commerce, NOAA, National Marine Fisheries Service, Southeast Regional Office. July. (Available at http://www.gulfcouncil.org/Beta/GMFMCWeb/downloads/Amend%2026%20FSEIS%20072706.pdf.)

———. 2007. *Fisheries of the United States—2006.* (Accessed 12/6/07 from http://www.st.nmfs.gov/st1/fus/fus06/index.html.)

National Research Council (NRC). Committee to Review Individual Fishing Quotas. 1999. *Sharing the Fish: Toward a National Policy on Individual Fishing Quotas.* Washington, DC: National Academies Press.

North Pacific Fishery Management Council (NPFMC). 2001. *Environmental Assessment/Regulatory Impact Review/Initial Regulatory Flexibility Analysis for a Regulatory Amendment to Incorporate the Halibut Charter Sector into the Halibut Individual Fishing Quota Program or Implement a Moratorium on Entry into the Charter Fleet for Pacific Halibut in Areas 2C and 3A.* Anchorage, AK: North Pacific Fishery Management Council, March 12.

North Pacific Fishery Management Council (NPFMC), and Institute for Social and Economic Research (ISER). 1997. *Environmental Assessment / Regulatory Impact Review / Initial Regulatory Flexibility Analysis (EA/RIR/IRFA) for Proposed Regulatory Amendments to Implement Management Alternatives for the Guided Sport Fishery for Halibut Off Alaska.* Condensed Draft for Council and Public Review. North Pacific Fishery Management Council and the University of Alaska Institute for Social and Economic Research. Anchorage, AK: North Pacific Fishery Management Council.

North Pacific Fishery Management Council (NPFMC), National Marine Fisheries Service, Alaska Department of Fish and Game, and the International Pacific Halibut Commission. 1998. *Proposed Halibut Guideline Harvest Level (GHL) Management Measures Discussion Paper.* Anchorage, AK: NPFMC.

Organisation for Economic Co-operation and Development (OECD). 1997. *Towards Sustainable Fisheries: Economic Aspects of the Management of Living Marine Resources.* Paris: OECD.

Pearse, Peter. 1991. *Building on Progress: Fisheries Policy Development in New Zealand.* Report prepared for the Minister of Fisheries. Wellington, NZ: Ministry of Fisheries, July.

Pomeroy, R. 1999. *Devolution and Fisheries Co-management.* Paper presented at the Workshop on Collective Action, Property Rights, and Devolution of Natural Resource Management, Puerto Azul, Philippines, June 21–25.

Schirripa, Michael J. 1998. Status of Red Snapper in the U.S. Waters of the Gulf of Mexico. *Contribution: SFD* 97/98-30. Miami, FL: NOAA, National Marine Fisheries Service, Southeast Fisheries Science Center.

Schirripa, M. J., and Christopher M. Legault. 1997. Status of the Red Snapper in U.S. Waters of the Gulf of Mexico: Updated through 1996. *Contribution: MIA* 97/98-05. Miami, FL: NOAA, National Marine Fisheries Service, Southeast Fisheries Science Center.

———. 1999. Status of the Red Snapper in U.S. Waters of the Gulf of Mexico: Updated through 1998. *Contribution: SFD*99/00-75. Miami, FL: NOAA, National Marine Fisheries Service, Southeast Fisheries Science Center.

Shotton, Ross, ed. 2001. Case Studies on the Allocation of Transferable Quota Rights in Fisheries. *FAO Fisheries Technical Paper,* No. 411. Rome: Food and Agriculture Organization of the United Nations.

Sissenwine, Michael P., and Andrew A. Rosenberg. 1993. Marine Fisheries at a Critical Juncture. *Fisheries* 18(10): 6–14.

Sissenwine, Michael P., and Pamela M. Mace. 2001. Governance for Responsible Fisheries: An Ecosystem Approach. Paper presented to the Reykjavik Conference on Responsible Fisheries in the Marine Ecosystem, Reykjavik, Iceland, October 1–4.

Sutinen, Jon G. 1993. Recreational and Commercial Fisheries Allocation with Costly Enforcement. *American Journal of Agricultural Economics.* 75(5): 1183–87.

———. 1996a. *Fisheries Compliance and Management: Assessing Performance.* Canberra: Australian Fisheries Management Authority, August.

———. 1996b. *Recreational Entitlements: Integrating Recreational Fisheries into New Zealand's Quota Management System.* Wellington: NZ Ministry of Fisheries, August.

Sutinen, Jon G., and Robert J. Johnston. 2003. Angling Management Organizations: Integrating the Recreational Sector into Fishery Management. *Marine Policy* 27(November): 471–87.

Sutinen, Jon G., Robert J. Johnston, and Reena Shaw. 2002. A Review of Recreational Fisheries in the U.S.: Implication of Rights-Based Fisheries. Department of Environmental and Natural Resource Economics, University of Rhode Island.

U.S. Fish and Wildlife Service (USFWS). 2002. *2001 National Survey of Fishing, Hunting, and Wildlife-Associated Recreation: National Overview.* Preliminary Findings. Washington, DC: U.S. Department of the Interior, Fish and Wildlife Service, May. (Available at http://library.fws.gov/nat_survey2001.pdf.)

Viswanathan, K. Kuperan, Nik Mustapha R. Abdullah, and Robert S. Pomeroy. 1998. Transaction Costs and Fisheries Co-Management. *Marine Resource Economics* 13: 103–14.

Index

aboriginal/native peoples, 31, 32, 33, 34, 54, 89, 91, 101
Alaska charter ITQ program: catch data, 156; demise of, 56, 134, 142; description of, 55–56, 132–34; economic analysis of, 134–39; program design considerations, 59, 63, 66, 67, 69, 131, 165n4, 193, 215, 226n11. *See also* Alaska halibut fishery
Alaska halibut fishery, ix–x, 10, 55–56, 132–34, 142n4, 156. *See also* Alaska charter ITQ program
allocation issues: court cases on, 33–35, 37–39; demand greater than supply, 182–83; harvest impacts and, 14, 182; harvest tag programs and, 188–90, 196n8; individual transferable fishing quotas (IQs, IFQs, or ITQs), 106–8, 139; management alternatives, *184*, 222, 223; market equilibrium and, *86*; in multiple-use fisheries, 5, 24, *38*, 79–94, 126, *136*, 141–42; optimality and, 24, 80–83, *81–82*, 87, 92, *107*, 129, 146–47; preferences to one sector, 27, 34, 135–36; program design considerations and, 64–66, 138–39, 188–89, 206–7; in sport fisheries, 163–64; unit determination, 129, 163–64. *See also* initial allocations
Amateur Fishing Trust, 42
angling management organizations (AMOs): access rights, 214; alternatives to, 221–24; costs during transition period, 220; description of, 67, 210–11, 221, 225; funding and sustainability of, 215–16; implementation issues, 217–24; initial allocations, 217–18; management considerations, 214, 219, 220–21; membership in, 213–14; monitoring and enforcement, 219–20; quota/share ownership, 212–13; quota trading, 215; quota units, 214–15; rights and duties, 212; scope and size, 216–17, 218–19. *See also* clubs
annual catch entitlement (ACE), 24–26
asset units, 52–54, 56–57, 59–61, 63, 70–71, 214–15
Atlantic States Marine Fisheries Commission, 208

— 231 —

auctions, 65, 71–72, 130, 178, *184*, 188–89, 192, 212
Australia pink snapper fishery, *180–81*, 182, 183, 186, 188

bag/catch limits: and allocations, 27; bioeconomic modeling, *51*; effectiveness of, 4, 33, 63; and mortality, 49–51; program design considerations and, 24, 59–60, 71, 156; red snapper, *48*, 204; reductions in, 31, 36, 165n4, 171, 173; slot limits, 179, 187; and sportfishing/charter boat fisheries, ix–x, 56, 155–58; as traditional management approach, 49–51
billfish, 179, 182
bioeconomic modeling, 24–26, 49–51, *51*, 80–88, 101–13, 116–20, 134–39, 146–47
biomass. *See* stock levels
black rockfish, 15–16, 17
blue cod, 23, 24
bluefish, 202
bycatch, 14, 16, 25, 73, 133, 143n5, 203–4, 226n15. *See also* mortality

catch limits. *See* bag/catch limits
Challenger Scallop Enhancement Company (CSEC), 40, 42
change, institutional, 7–13, 16–17
charter boat industry. *See* sportfishing/charter boat fisheries
Chinook salmon, x, 14–15, 16. *See also* salmon
Clean Air Act of 1990, 53
clubs, 28, 41–43, 44. *See also* angling management organizations (AMOs)
cod, 23, 24, *180*, 183, 188, 191
Columbia River salmon fishery, 11, 12, 14–15, 16, 114n2, *181*, 187
command-and-control management, 48–49, 52, 187
commercial fisheries: alternative management systems and, *137*; equity concerns, 13; individual transferable fishing quotas (IQs, IFQs, or ITQs), 54–55, 104–8, 126–29, 136; New Zealand, 23–26; and optimal allocation, *81–82*; Pacific Northwest, 9–17; and sportfishing/charter boat fisheries, 91–92, 124–39, *135*, *136*, 141–42, 146–47. *See also* multiple-use fisheries
common law, 25, 38
common/public property, fisheries as, 99, 130, 193, 225–26n4
the commons, x, 52
conflicts in multiple-use fisheries. *See* sector conflicts
Conservation Act 1987 (NZ), 42
conservation sector, 89–91, 93, 99–101, 103, 113, 179, 202

data collection: from charter boats, 30; management alternatives, 90, 179, 183, *185*, 191, 192; phone-and-diary surveys, 30, 32
data issues: costs, 206; management decisions affected by, 134, 191; monitoring of programs, 58, 71, 190; program design deficiencies, 32, 43; scarcity of recreational data, 6, 26, 29, 36, 67, 71; value measurements, 31, 84
Department of Fisheries and Oceans (Canada), 92, 94n1
derby fishing, 10, 12, 18n4
Duck Stamp Program, 175, *177*, 178, 192, 196n6
Dungeness crab, *181*, 187

economic modeling. *See* bioeconomic modeling
electronic tracking systems, 58, 61, 62, 68, 70, 71
Endangered Species Act (ESA), 5, 14, 130
enforcement. *See* monitoring and enforcement

equity concerns: in commercial fisheries, 13; and initial allocations, 71, 130, 217–18, 225–26n4; and maximum access, 17; and politics, 69
European Commission, 208
Exclusive Economic Zone (EEZ), 189
Executive Order 12962 (1995), 5

Federal Duck Stamp Program, 175, *177*, 178, 192, 196n6
fisheries: Alaska halibut, ix–x, 10, 55–56, 132–34, 142n4, 156; Australia pink snapper, *180–81*, 182, 183, 186, 188; Columbia River salmon, 11, 12, 14–15, 16, 114n2, *181*, 187; Florida tarpon, 179, *180–81*, 182, 183, 192; Gulf of Mexico grouper, 172, 173–74, 183, 186, 188, 194–95, 196n2, 198n40; Gulf of Mexico reef fish, 17, 172–73, 183–93, 198n40; Ireland salmon, 193; Ireland salmon and sea trout, *180–81*, 182, 183; Klamath River Chinook salmon, 16; Maryland and North Carolina multispecies, *180*, 182; Maryland multispecies, 179; Mid-Atlantic, 202; Newfoundland cod, *180*, 183, 188, 191; New Zealand kahawai, 23–24, 35–39, *37*, 41; New Zealand scallop, 39–40, *40*, 42; New Zealand snapper, 31–35, *32*, 41; North Atlantic salmon, 98–101; Oregon black rockfish, 15–16, 17; Pacific groundfish, 5, 6, 9–10, 11, 12, 13, 16, 18n3; Pacific halibut (Canada), 91–93, 94n1, 166n8; Scotland demersal, 54–55; South Dakota paddlefish, *180*, 182, 183, 188; summer flounder, ix. *See also* Gulf of Mexico red snapper fishery
Fisheries Act of 1983 (NZ), 26, 34
Fisheries Act of 1996 (NZ), 33, 34, 35, 37
Fishery Conservation and Management Act of 1976, 9–10, 130, 132–33

fixed gear, 10, 11–13, 15, 18n5
Florida Lifetime License, 197n37
Florida tarpon fishery, 179, *180–81*, 182, 183, 192

game hunting programs. *See* harvest tag programs
General Bioeconomic Fisheries Simulation Model (GBFSM), 50–52, *51*
gill nets, 11, 100–101
governments (national): and allocation issues, 83–86, 99; authority over fisheries, 44, 52, 66, 68, 85, 224, 225–26n4; and co-management, 28, 40, 94n1, 132–33, 193–94, 197n37, 208–9, 226n5; and conservation of fisheries, 9–10, 11; devolution of authority, 208–9, 210–11, 226n5; federal council system, 203; funding/revenue mechanisms, 26, 28, 29, 30, 64, 66, 67, 68, 209, 213; ownership of salmon rivers, 114n3; regulatory policies of, 16, 203, 225n2
groundfish: Newfoundland cod fishery, 183, 188; Pacific groundfish fishery, 5, 6, 9–10, 11, 12, 13, 16, 18n3
grouper, 172, 173–74, 183, 186, 188, 194–95, 196n2, 198n40. *See also* Gulf of Mexico reef fish fishery
guideline harvest level (GHL), 133–34, 161
Gulf of Mexico Fishery Management Council (GMFMC), 66, 204, 210, 220, 224
Gulf of Mexico grouper fishery, 172, 173–74, 183, 186, 188, 194–95, 196n2, 198n40. *See also* Gulf of Mexico reef fish fishery
Gulf of Mexico red snapper fishery: harvest statistics, ix, 186, *204*; program design considerations, 61–62, 65, 70–72, 142n3, 210, 226n7; restrictions, 49–*50*, *205*; status and

trends, 173, 203–5. *See also* Gulf of Mexico reef fish fishery
Gulf of Mexico reef fish fishery, 17, 172–73, 183–93, 198n40. *See also* Gulf of Mexico grouper fishery; Gulf of Mexico red snapper fishery
Gulf States Marine Fisheries Commission, 226n14

halibut: Alaska halibut fishery, ix–x, 10, 55–56, 132–34, 142n4, 156; bundled harvest tag programs, *181*; Pacific halibut fishery (Canada), 91–93, 94n1, 166n8; revenue statistics, 142n4; and trawl gear, 18n3. *See also* Alaska charter ITQ program
Halibut Convention of 1923, 132
harvest limits and quotas: Alaska halibut fishery, 133–34; Gulf of Mexico reef fish fishery, *48*, 186–87; and harvest levels, *204*; management alternatives, *184*; New Zealand kahawai fishery, 36; New Zealand scallop fishery, 39; Oregon black rockfish fishery, 15; Pacific halibut, 91
harvest tag programs: advantages, 183–93; bundled, 194–95, 198n40; challenges, 183; characteristics, *176–77*, *180–81*; effectiveness, 183; goals and objectives of, 175, 179, 182–83, 195; Gulf of Mexico reef fish fishery, 183–93; in hunting, 56–58, 59, 61, 65, 172–73, 174–79, *176–77*, 196n8; monitoring, enforcement, and compliance, 190–91; revenue generation, 192; sector integration, 192–93; and total allowable catch (TAC), 197n15; transferability, 189
hunting programs. *See* harvest tag programs

Idaho Sportsman's Package program, 194
Illinois Central R.R. Co. v. Illinois, 130

individual transferable fishing quotas (IQs, IFQs, or ITQs): advantages of, 90–91, 139–40, 171; anticipatory behaviors and incentives, 55, 142, 158, 162–63; asset unit for, 52; bioeconomic modeling, 24–26, 49–51, 80–88, 101–13, 116–20, 134–39, 146–47; in commercial fisheries, 54–55, 104–8, 126–29; compared to angling management organizations (AMOs), 210–11, 223–24, 226n15; compared to traditional management, 54; definition, 106, 126; integrated program designs, 108–11, 129–34; New Zealand, 23–26, 41, 44; in Pacific Northwest fisheries, 10–11, 12; and rents, 97; resistance to, 140, 208; in sport fisheries, 129–32, 158–64, *159*, *161*; transfers among programs/sectors, 111–13, 166n8. *See also* Alaska charter ITQ program; transferable rights (TRs) programs
initial allocations: auctions, 65, 71–72, 130, 178, *184*, 188–89, 192, 212; and equity concerns, 71, 130, 217–18, 225–26n4; fixed price licenses, 66; fixed- vs. percentage-based, 129; grandfathering, 64, 67, 71; importance of, 84–85, 106–7, 217–18, 225n4; lotteries (in fishing programs), 130, *180*, 182, *184*, 188–90, 212–13, 218; in multiple-use fisheries, 30, 38–39, 192–93, 207; options for, 64–70, 71–72, 218, 223; and program success/failure, 56; and transaction costs, 86–87, 88. *See also* allocation issues
institutional change, 7–13, 16–17
integrated management, principles of, 205–9, *206*
International Fisheries Commission, 132
International Pacific Halibut Commission (IPHC), 132–33, 165n4

Ireland salmon fishery, *180–81*, 182, 183, 193
Ireland sea trout fishery, *180*, 182

kahawai, 23–24, 35–39, *37*, 41
Kansas nonresident hunting program, 57–58, 59, 61, 63, 65, 68–69, 70, 190
Klamath River Chinook salmon fishery, 16

license limitation programs, 9–10, 11, 12, 13, 18n3, 39
litigation, 33–37, *38*
lotteries: in fishing programs, 130, *180*, 182, *184*, 188–90, 212–13, 218; in hunting programs, 57–58, 65, *176–77*, 178, 179

Malcolmson v. O'Dea, 25
management: alternative systems of, *137*, *184–85*, 221–24; balancing uses, 37–39, 202; and bioeconomic modeling, 24, 102; challenges, 174, 190–91; command-and-control, 48–49, 52, 187; cooperative, 40, 44, 67–68, 132–33, 186, 193–94, 210–11, 222–23, 226n5; costs, 187, 192, 209, 220; data delays, 134; goals and objectives of, 5, 134, 205; history of, 3–7; integrated, 205–9, *206*; new approaches to, 16, 41–43; sportfishing alternatives, 124–32; support for change, 13, 17; time demands of intersectoral conflicts, 129; traditional approaches to, x, 48–52, 54–55, 63, 173; trends in, 201, 202–3, 207
Maori, 31, 32, 33, 34, 54
Marine Recreational Fisheries Statistics Survey (MRFSS), 5, 73n1
marlin, *180*
Maryland multispecies fishery, 179, 182
maximum sustainable yield (MSY): alternative management systems and, *137*; controversies over, 33, 35, 38, 41; and economics, 25, 40, 135, *137*; harvest data, 32, *137*; program design considerations, 25, 30, 36, 62, 138
Mid-Atlantic recreational fisheries, 202
Minister/Ministry of Fisheries (NZ), 25–26, 28–29, 31–43
monitoring and enforcement: in angling management organizations (AMOs), 67, 209, 212, 216, 219–20, 224; in co-managed fisheries, 67–68; costs, 8, 12, 13, 42, 56, 67, 85–86, 113, 140, 187, 206, 223–24; electronic tracking systems, 58, 70, 72; in harvest tag programs, 60, 172, *184*, 190–91; incentives for compliance, 54, 63–64, 90, 163, 190–91, 211, 219–20; and mortality, 163; in multiple-use fisheries, ix–x, 211; and program design considerations, 59–60, 63–64; and quotas, 186; and quota units, 214–15; in rights-based systems, 44, 53, 56, 60–61, 67, 68, 70–71, 74n7, 106; in sportfishing/charter boat fisheries, 67, 156–57; in traditional management systems, 4, 13, 32, *184*
monitoring of programs: management alternatives, *184*; program design considerations, 68, 71, 72, 74n7
mortality: bioeconomic modeling, *51*; data issues, 74n2; in Gulf of Mexico red snapper fishery, 186, 203–4, 210; and harvest tag programs, 187; program design considerations, 59–60, 73, 206, 207, 226n7; and restrictions, 14, 49–52, 133, 156, 158, 165n1, 165n3, 171, 174. *See also* bycatch
multiple-use fisheries: biomass and harvest, *105*; litigation, 33–35, 37–39, *38*, 202; monitoring and enforcement, ix–x; program design considerations, 24–30, 41–43, 66–68, 132–39, 206; separate property rights

236 *Index*

systems, 111–13; trades/transfers across sectors, 44, 92, *110*, *185*, 215. *See also* allocation issues; individual transferable fishing quotas (IQs, IFQs, or ITQs); sector conflicts

National Academy of Public Administration (NAPA), 202
National Marine Fisheries Service (NMFS), 4, 5, 66, 173–74, 202, 225n2
National Policy for Fisheries Management (NZ), 27
Newfoundland cod fishery, *180*, 183, 188, 191
New Zealand Big Game Fishing Council, 37
New Zealand kahawai fishery, 23–24, 35–39, *37*, 41
New Zealand Recreational Fishing Council (NZRFC), 28, 30, 32, 33, 37
New Zealand scallop fishery, 39–40, *40*, 42
New Zealand snapper fishery, 31–35, *32*, 41
New Zealand's quota management system. *See* quota management system (QMS)
North Atlantic salmon fishery, 98–101
North Atlantic Salmon Fund (NASF), 91, 99–101
North Carolina multispecies fishery, 179, *180*, 182
North Pacific Fishery Management Council (NPFMC), 56, 133–34, 142, 196n1

open access: alternative management systems for, 124–29, 136, *137*; economics of, 43, 79, 136, 158–61, 164–65; in New Zealand fisheries, 39, 44; in Pacific Northwest fisheries, 13, 18n3; regulated, 124–26, 155–58, 165n3; in sport fisheries, 124–28, 152–58, *153*

optimality, 24–25, 30, 41, 44n2, 80–84, 93, 116–20, 129, 146–47
Option 4 group, 29
Oregon black rockfish fishery, 15–16, 17
Oregon Combined Angling Harvest Tag program, 194
Oregon combined harvest programs, 53–54, *181*, 182, 183, 194
Oregon Fish and Wildlife Commission (OFWC), 15
Oregon Nearshore Strategy, 15
Oregon Sports Pac program, 194
organizations of recreational fishers, 91, 92, 99–101. *See also* angling management organizations (AMOs); clubs
overfishing/overfished areas, ix, x, 5, 26, 27, 173–74, 202, 203

Pacific Fishery Management Council (PFMC), 4, 10–11, 16, 18n1, 18n3
Pacific groundfish fishery, 5, 6, 9–10, 11, 12, 13, 16, 18n3
Pacific halibut fishery (Canada), 91–93, 94n1, 166n8
paddlefish, *180*, 182–83, 188
permit stacking, 11, 18n5
pink shrimp, 18n3
pink snapper, *180–81*, 182, 183, 186, 188
politics: and allocations, 7, 12, 71, 72, 129, 134, 139–41, 218; and economics, 8, 11, 13, 62, 69, 136, 138; and equity concerns, 13, 69, 221; and program design considerations, 6, 58, 62, 71–72; and quota systems, 16; and trading of rights, 17, 43, 69
pollution control trading programs, 53–54
property rights: of aboriginal/native peoples, 29; characteristics of, 11, 172; characteristics of rights-based management, 172; commercial and recreational comparisons, 17, 26; and common property issues, 99,

130, 193, 225–26n4; and institutional change, 7–13; interest group theory and, 8; and market systems, 79, 97; perfect vs. imperfect, 105–6; and resource stewardship, 163, 172–73, 190–91, 207, 209; separate vs. integrated systems of, 111–13
public good(s), 29, 41, 89, 113

Quota Appeal Authority (NZ), 31
quota management system (QMS): description of, 23–27; and kahawai fishery, 36, *37*, 38; and program design considerations, 43–44; and recreational fisheries, 31, 32, 42; and scallop fishery, 39–40; and snapper fishery, 31, 35
quotas. *See* harvest limits and quotas; individual transferable fishing quotas (IQs, IFQs, or ITQs)

race for/to fish, 18n4, 54, 55, 125, 128, 153, 165n1, 189, 202, 226n15
red drum, 179, 187, 202
red snapper: harvest quotas and statistics, ix, 186, *204*; program design considerations, 61–62, 65, 70–72, 198n40, 210, 226n7; restrictions, *49–50, 205*; status and trends, 173, 203–5. *See also* Gulf of Mexico reef fish fishery
Reef Fish Fishery Management Plan, 173, 210
regulations and restrictions: in Alaska halibut fishery, 59, 133–34, 142; in angling management organizations (AMOs), 214; in co-managed fisheries, 209, 222; commercial vs. recreational fisheries, 17, 44; compliance with, 191, 209; goals and objectives of, 3, 27–28, 49, 165n3; in Gulf of Mexico fisheries, *48*, 171, 173, 189, 204, 219; in harvest tag programs, 60, 174–75, 182, 183;

market methods vs., 83–84, 164; and open access, 155–58; optimal levels of, 6; in Pacific Northwest fisheries, 4, 10, 16; power struggles over, 7; and program design considerations, 62; protests about, 37; in rights-based systems, 54, 63, 65, 68–69, 94; in sportfishing/charter boat fisheries, 93, 134, 152, 155–56, 158, 164; status and trends, ix–x, 5, 47, 171; in traditional management systems, 11, 15, 25–26, 54–55, 171, *184–85*, 203; on transfers and trades, 69, 72, 85, 93, 139, 215; trends, 171, 203, 204; voluntary measures and, 90, 156–57, 162, 165, 190–91, 223–24. *See also* monitoring and enforcement; *individual types of restrictions*
rent(s): and allocations, 66, 108; and costs, 207; in derby fisheries, 12; dissipation of, 79, 129; in harvest tag programs, 195; in hunting programs, 57–58; and individual transferable fishing quotas (IQs, IFQs, or ITQs), 79, 97, 160; and maximum sustainable yield (MSY), 25; and politics, 13
rent-seeking behaviors, 8, 66, 193, 213, 226n9
resource stewardship: government responsibility for, 130, 202, 209; property rights and, 172–73, 190–91, 207, 209; revenue and, 163–64, 212; role of recreational fisheries in, 3, 225n3, 226n7
restrictions. *See* regulations and restrictions
ridgeback prawns, 18n3
rights-based management. *See* property rights
rockfish, 15–16, 17

sablefish, 10–11, 12, 13, 18n5, 143n5
sailfish, *180*

salmon: bundled harvest tag programs, *181*, 187, 194; Columbia River fisheries, 11, 12, 14–15, 16, 114n2, *181*, 187; Ireland fishery, *180–81*, 182, 183, 193; North Atlantic fishery, 98–101, *100*, 114n4; overfishing/overfished areas, 5; state ownership of, 114n3. *See also* Chinook salmon
scallop, 39–40, *40*, 42
Scotland demersal fishery, 54–55
scup, 202
sea cucumbers, 18n3
season length/closures: bioeconomic modeling, *51*; disadvantages of, 93, 187; and harvest tag programs, 187–88; management alternatives, *184*; monitoring and enforcement, 55; red snapper, *48*, *50*, 173, 187–88, 204, *205*; as traditional management approach, 49, 165n3; trends, 171
sea trout, *180*, 182
sector conflicts: allocation issues, 14–15, 202, 206–7; bioeconomic modeling of, 102–11; growth of recreational sector and, 133, 201, 202, 207, 225n3; rights-based management and, 90–91, 97–98, 113, 192–93, 216–17; value issues, 6, 138
Shark Bay fishery. *See* Australia pink snapper fishery
shrimp, 18n3, 203, 204
size limits: changes in, 47, *48*, 171, 210; conflicts over, ix; goals and outcomes of, x, 49, 71, 195; Gulf of Mexico policies on, *48*, 173–74, 186, 195, 204, 210; in harvest tag programs, 50, 182, *184–85*, 187, 191; and mortality, 60, 155, 158, 171; New Zealand policies on, 25, 26–27, 32, 34, *40*; Pacific recreational policies on, 4; and program design considerations, 49–51, 59, 60, 63, 71, 128, 156, 165, 191, 192
slot limits, 179, 187

snapper, 31–35, *32*. *See also* pink snapper; red snapper
Soundings, 28–29
South Dakota paddlefish fishery, *180*, 182, 183, 188
sportfishing/charter boat fisheries: Alaska charter ITQ program, 56; alternative management systems for, 124–32, *137*, 152–63, 216, 219; anticipatory behaviors and incentives, 156–57, *160*; catch limit reductions, ix–x; characterized as recreational, 44; and commercial fisheries, 91–92, 124–39, *135*, *136*, 141–42, 146–47; contributions to economy, 23; and data collection, 30, 157; description of, 142n1; future regulations, 44n1; individual transferable fishing quotas (IQs, IFQs, or ITQs), 129–32, 149, *159*, 224; monitoring and enforcement, 60; proportion of recreational catch, 4; resident vs. nonresident markets, 150–58, *153*, *157*; restrictions, 6, 128; trip demand, 124–25, 152; trip prices, 125, 128, 131, 135, 143n5, 151–52, *153*, *157*, 159–62, *161*
steelhead, *181*, 187, 194
stock levels: as capital assets, 213; data issues, 38; effects of changes in, 125–29, 132, 135, *136*, 143n7, 149, 154–55, 164, 201, 202; and harvest quotas, *32*, *105*, 124, *137*; and maximum sustainable yield (MSY), 30; political control of, 140; and sector conflicts, 140
summer flounder, ix, 202
Sustainable Fisheries Act of 1996, 11, 12
swordfish, *180*

tarpon, 179, *180–81*, 182, 183, 192
Title IV of the 1990 Clean Air Act, 53
total allowable catch (TAC): in Alaska halibut fishery, 133; in angling

management organizations
 (AMOs), 210–11, 217, 219, 225; in
 bioeconomic modeling, 105–12, 113,
 118, 152, 155, *157*; in commercial
 fisheries, 52, 138, 225n1; in Gulf of
 Mexico fisheries, *204*, 210, 225n1;
 increase in recreational share,
 94n1; in management councils and
 advisory committees, 222–23; in
 New Zealand fisheries, 25–26, 27,
 31, *32*, 33, 35–36, *36*, *40*, 41; politics
 and, 139; and restrictions, 54–55,
 157
total allowable commercial catch
 (TACC), 24–26, 31–33, 35, 36, 41
total allowable noncommercial catch
 (TANC), 25, *32*, 36
transaction costs: and allocations, 80,
 85–87, *86*, 88, *88*, 92; in angling
 management organizations (AMOs),
 211, 220, 225; and change, 8–9, 12;
 electronic tracking systems, 58, 68; in
 harvest tag programs, 60; in hunting
 programs, 58; in Pacific Northwest
 fisheries, 10, 11, 13, 15, 16–17; and
 program design considerations, 53,
 56, 59–61, 70, 93–94

transferable rights (TRs) programs: and
 allocation issues, 87–88; in hunting
 programs, 56–58, 59, 61; overview,
 52; in pollution control, 53–54;
 program design considerations,
 58–70, 197n18. *See also* individual
 transferable fishing quotas (IQs,
 IFQs, or ITQs)
trawl gear, 10, 13, 15, 18n3, 35–36, 203
trip limits, 10, 15, 18n2, 128
troll gear, 11, 16
tuna, 35, *180*

U.S. Migratory Bird Hunting and
 Conservation Stamp Program, 175,
 177, 178, 192, 196n6

value: differences among sectors, 89–91;
 difficulties of determining, 31, 84;
 open access and dissipation of, 43, 79;
 and optimal allocation, 83; trading of
 rights and, 72, 92–93
Vigfusson, Orri, 100

Washington catch record card program,
 181, 182, 183, 187
water quality trading programs, 53–54

About the Contributors

Ragnar Arnason is a professor of fisheries economics at the University of Iceland. With a master's degree in mathematical economics and econometrics from the London School of Economics, he received his Ph.D. in natural resource economics from the University of British Columbia in 1984. Since becoming a professor in fisheries economics in 1989, Professor Arnason has primarily conducted his research in the area of fisheries economics and fisheries management where he has a publication record of over 130 scientific articles and books. Professor Arnason has played an important role in the development of individual tradable transferable quotas (ITQs) for Icelandic fisheries. He has also been active in consultancy work around the world on fisheries policy and fisheries management systems.

Tammy Warner Campson is a Ph.D. student in the Department of Agricultural and Resource Economics at the University of Connecticut. Her research is focused on evaluating the effectiveness of alternative fisheries management regimes and institutions. She holds an undergraduate degree in biology from the University of California at Santa Cruz and an MBA from Cornell University. Her private-sector experience includes both corporate management and small business consulting.

Keith R. Criddle is the Ted Stevens Distinguished Professor of Marine Policy in the University of Alaska, Fairbanks' Juneau Center for Fisheries and Ocean Science. Dr. Criddle received his Ph.D. in agricultural economics from the University of California, Davis in 1989. Dr. Criddle's research focuses on the intersection between the natural sciences and economics, especially the management of living resources. Dr. Criddle's research has explored topics ranging from the economic consequences of alternative management regimes for the governance of commercial, sport, and subsistence fisheries to the bioeconomic effects of climate change in North Pacific fisheries. Dr. Criddle is a member of the National Academies' Ocean Studies Board, an associate editor of *Natural Resource Modeling*, and vice chair of the North Pacific Fishery Management Council's Scientific and Statistical Committee.

Wade L. Griffin is a professor of agricultural economics at Texas A&M University. He received his B.S. and M.S. from Texas Tech University and his Ph.D. from Oregon State University, all in agricultural economics. For the past thirty-five years, he has enjoyed a national and international reputation in policy analysis with respect to fisheries and aquaculture and has obtained more than $2.5 million in competitive research grants to support his research program. He is best known for the development of the General Bioeconomic Fisheries Simulation Model (GBFSM) for the purpose of evaluating alternative management policies proposed by the Gulf of Mexico Fisheries Management Council and the National Marine Fisheries Service.

Susan S. Hanna is professor of marine economics at Oregon State University, affiliated with the Coastal Oregon Marine Experiment Station and with Oregon Sea Grant. Her research and publications are in the area of marine economics and policy, with an emphasis on fishery management, ecosystem-based fishery management, property rights, and institutional design. Dr. Hanna has served as a scientific advisor to the U.S. Commission on Ocean Policy, National Oceanic and Atmospheric Administration, National Marine Fisheries Service, Minerals Management Service, Northwest Power and Conservation Council, and the Pacific Fishery Management Council. She has served on the Ocean Studies Board of the National Research Council (NRC) of the National Academy of Sciences and has served on several NRC Committees, including the Committee to Review Individual Quotas in Fisheries and the Committee on Protection and Management of Pacific Northwest Anadromous Salmonids.

Daniel S. Holland is a resource economist with the Gulf of Maine Research Institute (GMRI). Prior to joining GMRI, he held positions as senior economist for the New Zealand Seafood Industry Council, assistant professor at the University of Massachusetts Dartmouth, and industry economist for the Alaska Fisheries Science Center. He received his Ph.D. in environmental and natural resource economics from the University of Rhode Island in 1998. His research is focused on design and evaluation of fishery management tools and strategies that will lead to profitable and sustainable fisheries and a healthy marine ecosystem. His research methods include bioeconomic modeling, econometric analysis, and experimental economics. Dr. Holland actively participates in the development of fishery policy by working with fishery stakeholders and managers to develop and evaluate policy. He is an associate editor of the journal *Marine Resource Economics*.

Robert J. Johnston is director of the George Perkins Marsh Institute and professor of economics at Clark University. He holds a B.A. in economics from Williams College and a Ph.D. in environmental and natural resource economics from the University of Rhode Island. Dr. Johnston is a distinguished member and director of the Northeastern Agricultural and Resource Economics Association and vice president of the Marine Resource Economics Foundation. He serves on the editorial board of the *Agricultural and Resource Economics Review*. He is an internationally recognized expert in the valuation of nonmarket resources and ecosystem services, benefit transfer, management of coastal and marine resources, and tourism economics. Dr. Johnston has published more than fifty articles in scholarly journals such as the *Journal of Environmental Economics and Management*, *Land Economics*, and the *American Journal of Agricultural Economics*.

Hwa Nyeon Kim is a Korean economist dedicated to the fields of renewable resources and environmental economics. At Texas A&M University, he received a Ph.D. in 2007, majoring in agricultural economics. His dissertation contained an institutional framework and theoretical and empirical models to evaluate the usability of transferable rights for the management of a recreational fishery. As a resource and environmental economist, he writes about various environmental issues, including the distributional impact of recreational fishing fees and the environmental Kuznets curve. He is currently a research fellow at Samsung Economic Research Institute (SERI), the number one private think tank in Korea. Since joining SERI, his research interest has been mainly in agricultural trade policies and commodity markets. His articles and essays often appear in Korean newspapers and magazines.

Donald R. Leal is senior fellow and research director at the Property and Environment Research Center (PERC) in Bozeman, Montana. His research areas of interests include property rights in marine fisheries, U.S. western water markets, and economic institutions for inland fish and wildlife. He is coauthor with Terry Anderson of *Free Market Environmentalism* (1991), which received the 1992 Choice Outstanding Academic Book Award and the 1992 Sir Antony Fisher International Memorial Award. The revised edition of that book was published in 2001. He is also editor of *Evolving Property Rights in Marine Fisheries* (2004) and has published numerous articles on such topics as individual fishing quotas (IFQs) in ocean fisheries, water marketing for fish and wildlife, the creation of self-sustaining parks, and the application of the trust concept to public lands. He currently serves on the Ad Hoc Grouper IFQ Advisory Panel for the Gulf of Mexico Fishery Management Council. He received his B.S. in mathematics and M.S. in statistics from California State University at Hayward.

Vishwanie Maharaj is an economist with the Environmental Defense Fund in their Austin, Texas, office, where she works to advance economic and business strategies to restore the Gulf of Mexico's fishery and coastal resources. Projects include development of new market-based tools for fishery management and supervision of research projects to depict the economic and conservation benefits of IFQ programs. Prior to joining the Environmental Defense Fund, Dr. Maharaj served as the director of economics for the American Sportfishing Association and economist with the South Atlantic Fishery Management Council. Past research projects include studies to value recreational fisheries and studies to measure the economic impact of recreational fishing. Dr. Maharaj has also served as a member of the federal Ecosystems Approach and Federal Investment task forces to study a broader ecosystem context for fisheries management and the federal government's role in subsidizing capitalization in U.S. fisheries.

Peter H. Pearse is professor emeritus of economics and forestry at the University of British Columbia. In addition to his academic career, Dr. Pearse has conducted several public inquiries on natural resources in Canada, including Royal Commissions on British Columbia's forest resources and Canada's Pacific fisheries. Dr. Pearse has also served as an advisor on the management of forests and other natural resource issues to foreign governments, the World Bank, and other international organizations. He is now a consultant on natural resource and environmental issues. His extensive publications deal mainly with the management of forests, fisheries, water resources, and the natural environment. Among other distinctions, Dr. Pearse has been awarded the Canadian Forestry Achievement Award, the Distinguished Forester Award, the Queen's Golden Jubilee Medal, and the Order of Canada.

Basil M. H. Sharp is a professor of economics at the University of Auckland, New Zealand. His research has focused on the evolution of property rights in New Zealand's marine fisheries. Dr. Sharp's published papers include the introduction of a rights-based system of governance; the structure, functioning, and efficiency of quota markets; and technological change in the seafood sector. His recent work includes an analysis of New Zealand's catch balancing regime in multispecies fisheries and contemporary problems of allocating fish stocks in shared fisheries. Dr. Sharp has worked as an advisor to the Food and Agriculture Organization of the United Nations, the New Zealand Ministry of Fisheries, and the fishing industry.

Jon G. Sutinen is professor emeritus of the Department of Environmental and Natural Resource Economics at the University of Rhode Island. Dr. Sutinen earned his Ph.D. in economics in 1973 from the University of Washington. Dr. Sutinen is past president of the North American Association of Fisheries Economists (2003–2005); and he recently was chair of the National Research Council Committee on Defining Best Available Science (2003–2004), a member of the Ocean Studies Board (2001–2003), and co-chair of the Social Science Advisory Committee of the New England Fisheries Management Council. In addition, Dr. Sutinen was the founding editor of the journal *Marine Resource Economics* and served in that role for over a decade.

James E. Wilen is director of the Center for Natural Resource Policy Analysis and professor of agricultural and resource economics at the University of California, Davis. Previously, he was at the Universities of British Columbia and Washington. He received his B.A. degree from California State University, Sonoma in 1970 and Ph.D. from the University of California, Riverside in 1973. His research focuses broadly on natural resource economics. His specific interests include bioeconomic modeling, open-access exploitation, factor distortion under regulated open access, and spatial models of resource use. He has received research and teaching awards, including the American Agricultural Economics Association (AAEA) Quality of Research Award (1998, 2000, 2004), the Western Agricultural Economics Association Outstanding Published Research Award (1998), AAEA Distinguished Graduate Teaching Award (1998), the University of California, Davis, Graduate Mentor Award (2004), and Supervisor-AAEA Outstanding Dissertation Award (six times). He was elected Distinguished Fellow of the AAEA in 2001.

Richard T. Woodward is an associate professor in the Department of Agricultural Economics at Texas A&M. He received his master's and Ph.D. from the University of Wisconsin, Madison. He is the author or co-author of over twenty-five articles and coauthor of the text book *Introduction to Agricultural Economics*. His research has touched on many aspects of environmental and resource economics. Dr. Woodward's recent work has focused on applications of market mechanisms in fisheries, water pollution, and the protection of environmental services in Costa Rica.